Thermal Power Plants Handbook

Thermal Power Plants Handbook

Edited by **Matt Danson**

CLANRYE INTERNATIONAL

New Jersey

Published by Clanrye International,
55 Van Reypen Street,
Jersey City, NJ 07306, USA
www.clanryeinternational.com

Thermal Power Plants Handbook
Edited by Matt Danson

International Standard Book Number: 978-1-63240-494-7 (Hardback)

Printed in the United States of America.

Contents

Permissions

List of Contributors

Preface

Every book is initially just a concept; it takes months of research and hard work to give it the final shape in which the readers receive it. In its early stages, this book also went through rigorous reviewing. The notable contributions made by experts from across the globe were first molded into patterned chapters and then arranged in a sensibly sequential manner to bring out the best results.

A facility for the production of electrical energy from thermal energy released by combustion of a fuel or consumption of a fissionable material is known as a Thermal Power Plant. Thermal power plants are significant process industries for engineering specialists. The power sector has been facing several crucial issues over the past few years. The primary challenge is to meet the increasing power demand in a sustainable and efficient manner. Practicing power plant engineers not only look after the maintenance and operations of the plant, but also look after a variety of activities like research and development, starting from power generation to the environmental evaluation of the power plants. This book discusses features, operational matters, advantages and limitations of power plants, as well as the benefits of renewable power generation. It also elucidates thermal performance evaluation, fuel combustion matters, performance monitoring and modeling, component fault diagnosis and prognosis, functional analysis, economics of plant operation and maintenance, and environmental facets. This book discusses numerous issues related to both coal fired and gas turbine power plants. It will be beneficial for undergraduate and research oriented students, and for engineers working in power plants.

It has been my immense pleasure to be a part of this project and to contribute my years of learning in such a meaningful form. I would like to take this opportunity to thank all the people who have been associated with the completion of this book at any step.

Editor

Part 1

Green Power Generation, Performance Monitoring and Modelling

Process Performance Monitoring and Degradation Analysis

Liping Li

China Power Engineering Consulting (Group) Corporation
China

1. Introduction

The global power sector is facing a number of issues, but the most fundamental challenge is meeting the rapidly growing demand for energy services in a sustainable way. This challenge is further compounded by the today's volatile market - rising fuel costs, increased environmental regulations, etc. Plant owners are challenged to prepare for the impact of future fuel price increases and carbon taxes and consider the value of environmental stewardship. The increasing competition in the electricity sector has also had significant implications for plant operation, which requires thinking in strategic and technical ways at the same time.

Management focus in the past decade has been on reducing forced outage rates, with less attention paid to thermal performance. Energy-intensive facilities seeking to maximize plant performance and profitability recognize the critical importance of performance monitoring and optimization to their survival in a competitive world. It means getting more out of their machinery and facilities. This can be accomplished through effective heat rate monitoring and maintenance activities. At present, it becomes necessary to find an uncomplicated solution assisting thermal performance engineers in identifying and investigating the cause of megawatt (MW) losses as well as in proposing new ways to increase MW output.

In this field of research and engineering, traditional system performance test codes [1] conduct procedures for acceptance testing based on the fundamental principles of the First Law of Thermodynamics. Many scholars have devoted to exergy-based research for the thermoeconomic diagnosis of energy utility systems [2-8], that is, those approaches based on the Second Law of Thermodynamics. In addition, some artificial intelligence model based methods [9-11] are also investigated for the online performance monitoring of power plant. However, some shortcomings also exist for the three kinds of methodologies. As is well known, performance test codes need sufficient test conditions to be fulfilled. It is difficult for continuous online monitoring condition to satisfy such rigorous requirements. Many artificial intelligence based methods may work well on data extensive conditions, but can't explain the results explicitly. Exergy analysis is very valuable in locating the irreversibilities inside the processes, nevertheless it needs to be popularized among engineers.

In this chapter, a novel method is presented, which is deduced from the First Law of Thermodynamics and is very clear and comprehensible for maintenance engineers and operators to understand and make use of. It can also sufficiently complement test codes. The novelty mainly lies in as followings: first, the primary steam flow is calculated indirectly by existing plant measurements from system balance to alleviate test instrument installation and maintenance cost compared with standard procedure. Furthermore, the measurement error can be avoided instead of direct use of plant instrumentation for indication. Second, the degradation analysis technique proposed comes from the First Law of Thermodynamics and general system theory. It is very comprehensible for engineers to perform analysis calculation combining system topology that they are familiar with. Moreover, the calculated results from parameters deviation have traced the influences along the system structure beyond the traditional component balance calculation. Third, the matrix expression and vectors-based rules are fit for computer-based calculation and operation decision support.

2. Foundations for the new analysis method

In nature, the thermal system of a power unit is a non-linear, multi-variable and time-variant system. For the system performance analysis and process monitoring, two aspects are especially important.

First, process performance monitoring requires instrumentation of appropriate repeatability and accuracy to provide test measurements necessary to determine total plant performance indices. Available measurements set must be selected carefully for the proper expression of system inner characteristics. The benefits afforded by online performance monitoring are not obtained without careful selection of instrumentation. Moreover, calculated results are rarely measured directly. Instead, more basic parameters, such as temperature and pressure, are either measured or assigned and the required result is calculated as a function of these parameters. Errors in measurements and data acquisition are propagated into the uncertainty of the resulting answer. Measurement error should be considered combining with engineering availability and system feature itself.

Second, it's hardly possible to solve the hybrid dynamic equations consisting of fluid mechanics and heat transfer for a practical large physical system. From the point of view of system analysis [12], one nature system should be characterized by how many inputs and outputs they have, such as MISO (Multiple Inputs, Single Output) or MIMO (Multiple Inputs, Multiple Outputs), or by certain properties, such as linear or non-linear, time-invariant or time-variant, etc.

The following subsection briefly discusses two fundamental theories, which are employed to cope with issues mentioned above and support the new monitoring and analysis approach proposed in the chapter.

2.1 Measurement error and error propagation
2.1.1 General principles of error theory

Every measurement has error, which results in a difference between the measured value, X, and the true value. The difference between the measured value and the true value is the total error, δ. Total error consists of two components: random error ε and systematic error β. Systematic error is the portion of the total error that remains constant in repeated measurements throughout the conduct of a test. The total systematic error in a measurement is usually the sum of the contributions of several elemental systematic errors, which may arise from imperfect calibration corrections, measurement methods, data reduction techniques, etc. Random error is the portion of the total error that varies randomly in repeated measurements throughout the conduct of a test. The total random error in a measurement is usually the sum of the contributions of several elemental random error sources, which may arise from uncontrolled test conditions and nonrepeatabilities in the measurement system, measurement methods, environmental conditions, data reduction techniques, etc. In other words, if the nature of an elemental error is fixed over the duration of the defined measurement process, then the error contributes to the systematic uncertainty. If the error source tends to cause scatter in repeated observations of the defined measurement process, then the source contributes to the random uncertainty.

As far as error propagation is concerned, for a MISO system, $Y = f(X_1, X_2, \cdots, X_n)$, the effect of the propagation can be approximated by the Taylor series method. If we expand $f(X_1, X_2, \cdots, X_n)$ through a Taylor series in the neighborhood of $\mu_{X_1}, \mu_{X2}, \cdots, \mu_{X_n}$, we get

$$f(X_1, X_2, \cdots, X_n) = f(\mu_{X_1}, \mu_{X_2}, \cdots, \mu_{X_n}) + \theta_{X_1}(X_1 - \mu_{X_1}) + \theta_{X_2}(X_2 - \mu_{X_2}), \cdots,$$

$$+\theta_{X_n}(X_n - \mu_{X_n}) + \text{higher order terms} \tag{2.1}$$

Where θ_{X_i} are the sensitivity coefficients given by $\theta_{X_i} = \dfrac{\partial Y}{\partial X_i}$. Now, suppose that the arguments of the function $f(X_1, X_2, \cdots, X_n)$ are the random variables $X_1, X_2, \cdots, X_n$, which are all independent. Furthermore, assume that the higher order terms in the Taylor series expansion for $f(X_1, X_2, \cdots, X_n)$ are negligible compared to the first order terms. Then in the neighborhood of $\mu_{X_1}, \mu_{X_2}, \cdots, \mu_{X_n}$ we have

$$f(X_1, X_2, \cdots, X_n) = f(\mu_{X_1}, \mu_{X_2}, \cdots, \mu_{X_n}) + \theta_{X_1}(X_1 - \mu_{X_1}) + \theta_{X_2}(X_2 - \mu_{X_2}), \cdots, +\theta_{X_n}(X_n - \mu_{X_n}) \tag{2.2}$$

then,

$$\mu_Y = E(Y) \approx f(\mu_{X_1}, \mu_{X_2}, \cdots, \mu_{X_n}) \qquad \sigma_Y^2 \approx \sum_{i=1}^{n} \theta_{X_i}^2 \sigma_{X_i}^2$$

In the same way, the total variance associated with a measured variable, X_i, can be expressed as a combination of the variance associated with a fixed component and the variance associated with a random component of the total error in the measurement, that is,

$$\delta = \beta + \varepsilon \tag{2.3}$$

Assuming fixed errors to be independent of random errors and no correlation among the random errors, then the general form of the expression for determining the combined standard uncertainty of a result is the root-sum-square of both the systematic and the random standard uncertainty of the result. The following simple expression for the combined standard uncertainty of a result applies in many cases:

$$\sigma_Y^2 = \sum_{i=1}^{n} \theta_{X_i}^2 \sigma_{\varepsilon_i}^2 + \sum_{i=1}^{n} \theta_{X_i}^2 \sigma_{\beta_i}^2 \tag{2.4}$$

The general error propagation rules indicated in (2.4) shows clearly that both systematic error and random error is propagated to the calculated results, approximately proportioning to the partial derivative of each variable. However, θ is determined by system model and the role of the variable in the model. Thus, variables choice becomes one of key steps to control calculation uncertainty.

2.1.2 Characteristics of thermal system measurements

System performance index is calculated as a function of the measured variables and assigned parameters. The instrumentation employed to measure a variable have different required type, accuracy, redundancy, and handling depending upon the use of the measured variable and depending on how the measured variable affects the final result. For example, the standard test procedure requires very accurate determination of primary flow to the turbine. Those are used in calculations of test results are considered primary variables. However, the rigorous requirements make this type of element very expensive. This expense is easy to justify for acceptance testing or for an effective performance testing program, but is unaffordable for online routine monitoring.

In normal operation and monitoring, the primary flow element located in the condensate line is used for flow indication, which is the least accurate and is only installed to allow plant operators to know approximate flow rate. As is well known, fouling issues are the main difficulties for the site instrumentation. Remember that a 1% error in the primary flow to the turbine causes a 1% error in calculated turbine heat rate, that is, 100% error

Variables	Measurement Variation	Heat rate Deviation
Main steam temperature	1°F	+0.07%
Reheated steam (cold) temperature	1°F	-0.04%
Reheated steam (hot) temperature	1°F	+0.05%
Final feedwater temperature	1°F	+0.03~+0.04%
Condensate water temperature (deaerator inlet)	1°F	-0.01~-0.03%
Feedwater temperature (final high pressure heater inlet)	1°F	+0.02~0.04%
Feedwater temperature (first high pressure heater inlet)	1°F	-0.05~0.08%
Main steam pressure	1%	+0.02~0.04%
Reheated steam (cold) pressure	1%	-0.05~-0.08
Reheated steam (hot) pressure	1%	+0.08%
Leakage of High Pressure Cylinder gland steam [a]	1%	-0.0013%
Leakage of Intermediate Pressure Cylinder gland steam [a]	1%	-0.002%
Primary condensate flow[b]	1%	+1%
Power	1%	-1%

Note: a. The leakage quantity is compared with its rated value itself.
b. For comparing, the primary condensate flow is also included.

Table 1. The effect on heat rate uncertainty of variables measurements

propagation. It means the primary flow from existing plant instrument is no longer competent to fulfill the system performance calculation. Selecting new measurements set becomes an imperative for the function.

Fortunately, with the improvement of I&C technology and modernization of power plants, these auxiliary water/steam flow instruments are installed and well maintained, such as secondary flow elements, blow down, drain water, etc. The more existing plant instruments are employed to construct system state equation that lays the foundation for the new methodology proposed in the chapter.

Now, let's focus on the error propagation of these calculation-related measurements. Tab.1 shows the uncertainty propagation of some measurements from a typical larger subcritical power unit. It is revealed that the effect of auxiliary water/steam flow is insignificant compared with these primary variables. On the other hand, most small diameter lines have low choked flow limits; therefore, the maximum flow scenario most likely has a small effect on heat rate.

In a word, a larger measurements set (here, refers to employing more existing plant instrument measurements) and their low uncertainty propagation property are the foundation of system-state-equation-based process performance calculation and system analysis.

2.2 State space modeling
2.2.1 State space model

The state space model of a continuous-time dynamic system can be derived from the system model given in the time domain by a differential equation representation. Consider a general nth-order model of a dynamic system represented by an nth-order differential equation:

$$\frac{d^n y(t)}{dt^n} + a_{n-1} \frac{d^{n-1} y(t)}{dt^{n-1}} + \cdots + a_1 \frac{dy(t)}{dt} + a_0 y(t)$$

$$= b_n \frac{d^n u(t)}{dt^n} + b_{n-1} \frac{d^{n-1} u(t)}{dt^{n-1}} \cdots + b_1 \frac{du(t)}{dt} + b_0 u(t)$$

For simplicity, it is presented for the case when no derivatives with respect to the input, that is:

$$\frac{d^n y(t)}{dt^n} + a_{n-1} \frac{d^{n-1} y(t)}{dt^{n-1}} + \cdots + a_1 \frac{dy(t)}{dt} + a_0 y(t) = u(t) \tag{2.5}$$

Then, if the output derivatives are defined as:

$$x_1(t) = y(t)$$

$$x_2(t) = \frac{dy(t)}{dt} = \overset{\bullet}{x}_1$$

$$\vdots \tag{2.6}$$

$$x_n(t) = \frac{dy^{n-1}(t)}{dt^{n-1}} = \overset{\bullet}{x}_{n-1}$$

$$\frac{dy^n(t)}{dt^n} = \overset{\bullet}{x}_n$$

Then , (2.5) can be expressed by matrix form as:

$$
\begin{bmatrix} \dot{x_1} \\ \dot{x_2} \\ \vdots \\ \dot{x}_{n-1} \\ \dot{x_n} \end{bmatrix} = \begin{bmatrix} 0 & 1 & 0 & \cdots & 0 \\ 0 & 0 & 1 & \cdots & \vdots \\ \vdots & \vdots & \vdots & \vdots & 0 \\ 0 & 0 & \cdots & \vdots & 1 \\ -a_0 & -a_1 & -a_2 & \cdots & -a_{n-1} \end{bmatrix} \begin{bmatrix} x_1(t) \\ x_2(t) \\ \vdots \\ x_{n-1}(t) \\ x_n(t) \end{bmatrix} + \begin{bmatrix} 0 \\ 0 \\ \vdots \\ 0 \\ 1 \end{bmatrix} u(t)
\tag{2.7}
$$

$$
y(t) = \begin{bmatrix} 1 & 0 & \cdots & 0 & 0 \end{bmatrix} \begin{bmatrix} x_1(t) \\ x_2(t) \\ \vdots \\ x_{n-1}(t) \\ x_n(t) \end{bmatrix}
\tag{2.8}
$$

Generally, for a MIMO system, the vector matrix expression is given by:

$$
\dot{x} = Ax + Bu
\tag{2.9}
$$

$$
y = Cx + Du
\tag{2.10}
$$

Where (2.9) is known as the state equation and (2.10) is referred to as the output equation; x is called the "state vector"; y is called the "output vector"; u is called the "input vector"; A is the "state matrix"; B is the "input matrix"; C is the "output matrix" and D is the "feed-forward matrix".

System structure properties and inner characteristic are indicated within the matrixes A , B , C and D, which can also be generally called property matrix. From mathematical perspective, the matrix equations are more convenient for computer simulation than an nth order input-output differential equation. But there are many advantages with combining system topology when they are used for system analysis.

2.2.2 System analysis assumption vs. system state space model

Thermodynamics is the only discipline theory to depend on to evaluate the performance of a thermal physical system. Nowadays, balance condition thermodynamics is usually employed for the analysis of such an actual industrial physics system [13]. It mainly comes from as followings:

First, at the steady states, an energy system has the least entropy production, i.e. the lowest energy consumption from Prigogine's minimum entropy production principle. Because there exist many energy storage components in such a complex energy system, it is almost meaningless to assess energy consumption rate under system dynamics.

Second, for such a continuous production system, the actual process with stable condition is much longer than its dynamic process from an engineering perspective. Each stable production process can be regarded as a steady state system. System performance assessment should be conducted at steady-states, which can also transfer from one steady state to another one for responding to production demand.

With the thermodynamic balance condition assumption, a mass and energy balance model can be conducted for each component at steady state of the system. Then the system state equation can be obtained through proper mathematic arrangements guided by system topology. The system state equation is composed of system thermodynamics properties and some auxiliary flows. By comparing the system state equation with the general form of system state space model, a vector based analysis approach is inspired under the required assumption, that is, a linear time-invariant system at steady state.

3. Steam-water distribution equation

The steam-water distribution standard equation for thermo-system of a coal-fired power plant is deduced basing on components balance under the First Law of Thermodynamics.

3.1 The steam-water distribution equation for a typical thermal system

A fictitious system with all possible types of auxiliary system configuration is shown in Fig.1. The dashed beside each heater is used to indicate the boundary of heater unit, which play an important role in the ascertainment of feedwater's inlet and outlet enthalpy. Note that the boundary for the extraction steam of each heater is the immediate extraction pipe outlet of turbine, that is, any auxiliary steams input/output from the main extraction steam

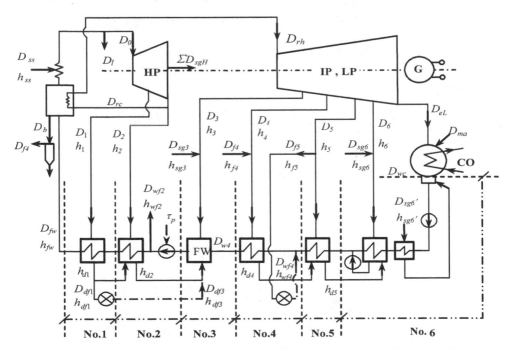

Fig. 1. Typical structure of thermal system (HP: high pressure cylinder; IP: intermediate pressure cylinder; LP: low pressure cylinder; CO: condenser ;G: generator; FW: deaerator)

pipe should be included in the respective heater unit (here, the term 'heater unit' is claimed to refer to the system control volume of heater defined on the above boundary rules.). Conducting mass balance and energy balance for each heater unit as the followings:

No.1:

$$D_1 h_1 + D_{fw} h_{w2} = D_1 h_{d1} + D_{fw} h_{w1}$$

$$\Rightarrow D_1 (h_1 - h_{d1}) = D_{fw}(h_{w1} - h_{w2})$$

$$\Rightarrow (D_1 - D_{df1})q_1 - D_{df1}(h_{df1} - h_1) = D_{fw}\tau_1 \tag{3.1}$$

No.2:

$$(D_1 - D_{df1})h_{d1} + D_2 h_2 + (D_{fw} + D_{wf2})(h_{w3} + \tau_p) = D_{fw}h_{w2} + [(D_1 - D_{df1}) + D_2]h_{d2} + D_{wf2}h_{wf2}$$

$$\Rightarrow (D_1 - D_{df1})\gamma_2 + D_2 q_2 + (D_{fw} + D_{wf2})\tau_p - D_{wf2}(h_{wf2} - h_{w3}) = D_{fw}\tau_2 \tag{3.2}$$

Where $h_{wf2} = h_{w3} + \tau_p$

No.3:

$$D_3 h_3 + D_{sg3}h_{sg3} + [(D_1 - D_{df1}) + D_2]h_{d2} + D_{df3}h_{df3} + D_{w4}h_{w4} = (D_{fw} + D_{wf2})h_{w3}$$

$$\Rightarrow (D_1 - D_{df1})\gamma_3 + D_2\gamma_3 + (D_3 + D_{sg3} + D_{df3})q_3 + D_{df3}(h_{df3} - h_3) + D_{sg3}(h_{sg3} - h_3) = (D_{fw} + D_{wf2})\tau_3 \tag{3.3}$$

where $D_{w4} = (D_{fw} + D_{wf2}) - (D_1 - D_{df1} + D_2 + D_{df3} + D_3 + D_{sg3})$
$h_{df3} = (h_{df3} - h_3) + h_3$, $h_{sg3} = (h_{sg3} - h_3) + h_3$

No.4:

$$(D_{w4} - D_{wf4})h_{w5} + D_4 h_4 + D_{f4}h_{f4} + D_{wf4}h_{wf4} = D_{w4}h_{w4} + (D_4 + D_{f4})h_{d4}$$

$$\Rightarrow (D_1 - D_{df1})\tau_4 + D_2\tau_4 + (D_3 + D_{sg3} + D_{df3})\tau_4 + (D_4 + D_{f4})q_4 + D_{f4}(h_{f4} - h_4) + D_{wf4}(h_{wf4} - h_{w5}) = (D_{fw} + D_{wf2})\tau_4 \tag{3.4}$$

Where $h_{f4} = (h_{f4} - h_4) + h_4$

No.5:

$$(D_{w4} - D_{wf4})h_{w6} + (D_4 + D_{f4})h_{d4} + (D_5 - D_{f5})h_5 = (D_{w4} - D_{wf4})h_{w5} + (D_4 + D_{f4} + D_5 - D_{f5})h_{d5}$$

$$\Rightarrow (D_1 - D_{df1})\tau_5 + D_2\tau_5 + (D_3 + D_{sg3} + D_{df3})\tau_5 + (D_4 + D_{f4})\gamma_5 + (D_5 - D_{f5})q_5 = (D_{fw} + D_{wf2} - D_{wf4})\tau_5 \tag{3.5}$$

No.6:

$$D_{uc}h_{uc} + D_6 h_6 + D_{sg6}h_{sg6} + D_{sg6}h_{sg6} + (D_4 + D_{f4} + D_5 - D_{f5})h_{d5} = (D_{w4} - D_{wf4})h_{w6}$$

$$\Rightarrow (D_1 - D_{df1})\tau_6 + D_2\tau_6 + (D_3 + D_{sg3} + D_{df3})\tau_6 + (D_4 + D_{f4})\gamma_6 + (D_5 - D_{f5})\gamma_6 +$$
$$(D_6 + D_{sg6})q_6 + D_{sg6}(h_{sg6} - h_6) + D_{sg6'}(h_{sg6'} - h_{wc}) = (D_{fw} + D_{wf2} - D_{wf4})\tau_6 \quad (3.6)$$

Where
$$D_{wc} = (D_{fw} + D_{wf2}) - (D_1 - D_{df1} + D_2 + D_{df3} + D_3 + D_{sg3}) - D_{wf4} -$$
$$(D_4 + D_{f4} + D_5 - D_{f5} + D_6 + D_{sg6} + D_{sg6'})$$

$$h_{sg6} = (h_{sg6} - h_6) + h_6$$

Rearranging (3.1) to (3.6) to matrix equation as (3.7) :

$$
\begin{bmatrix}
q_1 & & & & & \\
\gamma_2 & q_2 & & & & \\
\gamma_3 & \gamma_3 & q_3 & & & \\
\tau_4 & \tau_4 & \tau_4 & q_4 & & \\
\tau_5 & \tau_5 & \tau_5 & \gamma_5 & q_5 & \\
\tau_6 & \tau_6 & \tau_6 & \gamma_6 & \gamma_6 & q_6
\end{bmatrix}
\begin{bmatrix}
D_1 - D_{df1} \\
D_2 \\
D_3 + D_{sg3} + D_{df3} \\
D_4 + D_{f4} \\
D_5 - D_{f5} \\
D_6 + D_{sg6}
\end{bmatrix}
+
\begin{bmatrix}
-D_{df1}(h_{df1} - h_1) \\
(D_{fw} + D_{wf2})\tau_p - D_{wf2}(h_{wf2} - h_{u8}) \\
D_{df3}(h_{df3} - h_3) + D_{sg3}(h_{sg3} - h_3) \\
D_{wf4}(h_{wf4} - h_{w5}) + D_{f4}(h_{f4} - h_4) \\
0 \\
D_{sg6}(h_{sg6} - h_6) + D_{sg6'}(h_{sg6'} - h_{wc})
\end{bmatrix}
=
\begin{bmatrix}
\tau_1 D_{fw} \\
\tau_2 D_{fw} \\
\tau_3(D_{fw} + D_{uf2}) \\
\tau_4(D_{fw} + D_{wf2}) \\
\tau_5(D_{fw} + D_{wf2} - D_{wf4}) \\
\tau_6(D_{fw} + D_{wf2} - D_{wf4})
\end{bmatrix}
\quad (3.7)
$$

Substituting the following three equations to (3.7) and rearranging, we get (3.8):

$D_{fw} = D_0 + D_b + D_l - D_{ss}$ (from boiler flows balance)

$N_p = (D_{fw} + D_{wf2})\tau_p$ (the total enthalpy increase by feedwater pump shaft work)

$Q_{sg6'} = D_{sg6'}(h_{sg6'} - h_{wc})$ (the heat inputted by the turbine shaft gland steam heater)

In (3.8), $q_{df1} = h_{df1} - h_{d1}$; $q_{df3} = h_{df3} - h_{w4}$; $q_{sg3} = h_{sg3} - h_{w4}$; $q_{f4} = h_{f4} - h_{d4}$; $q_{sg6} = h_{sg6} - h_{wc}$; $q_{f5} = h_{f5} - h_{d5}$; $\tau_{wf2} = h_{wf2} - h_{w3}$; $\tau_{wf4} = h_{wf4} - h_{w5}$.

$$
\begin{bmatrix}
q_1 & & & & & \\
\gamma_2 & q_2 & & & & \\
\gamma_3 & \gamma_3 & q_3 & & & \\
\tau_4 & \tau_4 & \tau_4 & q_4 & & \\
\tau_5 & \tau_5 & \tau_5 & \gamma_5 & q_5 & \\
\tau_6 & \tau_6 & \tau_6 & \gamma_6 & \gamma_6 & q_6
\end{bmatrix}
\begin{pmatrix}
D_1 \\
D_2 \\
D_3 \\
D_4 \\
D_5 \\
D_6
\end{pmatrix}
+
\begin{bmatrix}
q_{df1} & & & & & \\
\gamma_2 & 0 & & & & \\
\gamma_3 & \gamma_3 & q_{sg3} & & & \\
\tau_4 & \tau_4 & \tau_4 & q_{f4} & & \\
\tau_5 & \tau_5 & \tau_5 & \gamma_5 & q_{f5} & \\
\tau_6 & \tau_6 & \tau_6 & \gamma_6 & \gamma_6 & q_{sg6}
\end{bmatrix}
\begin{pmatrix}
-D_{df1} \\
0 \\
D_{sg3} \\
D_{f4} \\
-D_{f5} \\
D_{sg6}
\end{pmatrix}
+
$$

$$
\begin{bmatrix}
0 & & & & & \\
\gamma_2 & 0 & & & & \\
\gamma_3 & \gamma_3 & q_{df3} & 0 & & \\
\tau_4 & \tau_4 & \tau_4 & 0 & & \\
\tau_5 & \tau_5 & \tau_5 & \gamma_5 & 0 & \\
\tau_6 & \tau_6 & \tau_6 & \gamma_6 & \gamma_6 & 0
\end{bmatrix}
\begin{pmatrix}
0 \\
0 \\
D_{df3} \\
0 \\
0 \\
0
\end{pmatrix}
+
\begin{bmatrix}
0 & & & & & \\
\tau_2 & \tau_{wf2} & & & & \\
\tau_3 & \tau_3 & 0 & & & \\
\tau_4 & \tau_4 & \tau_4 & \tau_{wf4} & & \\
\tau_5 & \tau_5 & \tau_5 & \tau_5 & 0 & \\
\tau_6 & \tau_6 & \tau_6 & \tau_6 & \tau_6 & 0
\end{bmatrix}
\begin{pmatrix}
0 \\
-D_{wf2} \\
0 \\
D_{wf4} \\
0 \\
0
\end{pmatrix}
+
$$

$$
-(D_b + D_l)
\begin{pmatrix}
\tau_1 \\
\tau_2 \\
\tau_3 \\
\tau_4 \\
\tau_5 \\
\tau_6
\end{pmatrix}
+ D_{ss}
\begin{pmatrix}
\tau_1 \\
\tau_2 \\
\tau_3 \\
\tau_4 \\
\tau_5 \\
\tau_6
\end{pmatrix}
+
\begin{pmatrix}
0 \\
N_p \\
0 \\
0 \\
0 \\
Q_{sg6'}
\end{pmatrix}
= D_0
\begin{pmatrix}
\tau_1 \\
\tau_2 \\
\tau_3 \\
\tau_4 \\
\tau_5 \\
\tau_6
\end{pmatrix}
\quad (3.8)
$$

For a more general system, i.e. there are r auxiliary steams and s auxiliary waters flowing in/out heater unit, t boiler blown down or other leakage flows. Then a general steam-water distribution equation can be got:

$$[\mathbf{A}]\,[\mathbf{D}_i]+\sum_{k=1}^{r}[\mathbf{A}_k]\,\left[\mathbf{D}_{fik}\right]+\sum_{j=1}^{s}\left[\mathbf{T}_j\right]\,\left[\mathbf{D}_{wfij}\right]-(\sum_{m=1}^{t}D_{bm})[\tau_i]+D_{ss}[\tau_i]+\left[\Delta\mathbf{Q}_{fi}^0\right]=D_0[\tau_i] \qquad (3.9)$$

Equation (3.9) can be written simply as :

$$[\mathbf{A}]\,[\mathbf{D}_i]+[\mathbf{Q}_{fi}]=D_0[\tau_i] \qquad (3.10)$$

Where,

$$[\mathbf{Q}_{fi}]=\sum_{k=1}^{r}[\mathbf{A}_k]\,\left[\mathbf{D}_{fik}\right]+\sum_{j=1}^{s}\left[\mathbf{T}_j\right]\,\left[\mathbf{D}_{wfij}\right]-(\sum_{m=1}^{t}D_{bm})[\tau_i]+D_{ss}[\tau_i]+\left[\Delta\mathbf{Q}_{fi}^0\right] \qquad (3.11)$$

$\left[\Delta\mathbf{Q}_{fi}^0\right]$ is the vector from pure heat imposed on the thermal system. For example, $\left[\Delta\mathbf{Q}_{fi}^0\right]=[0 \quad N_p \quad 0 \quad 0 \quad 0 \quad Q_{sg6}]$ in the demonstration system showed in Fig.1.

In (3.10), $[\mathbf{A}]$ is the character matrix consisting of the thermal exchange quantity (q,γ,τ) in each heater unit. $[\mathbf{D}_i]$ is vector consisting of the extracted steam quantity and $[\tau_i]$ is enthalpy increase of feeedwater (or condensed water) in each heater unit. D_0 is the throat flow of turbine inlet. $[\mathbf{Q}_{fi}]$ can be regarded as a equivalent vector consisting of the thermal exchange of all auxiliary steam (water) flow or external heat imposed on the main thermal system (here the main thermal system means the thermal system excluding any auxiliary steam or auxiliary water stream).

3.2 The transform and rearrangement of system state matrix equation of a general power unit

According to total differential equation transform, equation (3.12) can be obtained from equation (3.10), where the infinitesimal of higher order is neglected.

$$[\Delta\mathbf{A}][\mathbf{D}_i]+[\mathbf{A}][\Delta\mathbf{D}_i]+[\Delta\mathbf{Q}_{fi}]=D_0[\Delta\tau_i] \qquad (3.12)$$

Then,

$$[\Delta\mathbf{D}_i]=[\mathbf{A}]^{-1}\left[D_0[\Delta\tau_i]-[\Delta\mathbf{A}][\mathbf{D}_i]-[\Delta\mathbf{Q}_{fi}]\right] \qquad (3.13)$$

Considering the linear characteristics of the thermal system under a steady state, the system keeps its all components' performance constant while suffering the disturbance inputs coming from auxiliary steam (water) or external heat. That is, the thermal exchange of unit mass working substance in each heater unit is constant, then, the followings can be declared,

$$[\Delta\mathbf{A}]=0 , [\Delta\tau_i]=0$$

So equation (3.13) becomes,

$$[\Delta D_i] = -[A]^{-1}[\Delta Q_{fi}] \tag{3.14}$$

Doing total differential equation transform and neglecting the infinitesimal of higher order, and the equation (3.11) becomes,

$$[\Delta Q_{fi}] = \sum_{k=1}^{r}[\Delta A_k]\left[D_{fik}\right] + \sum_{k=1}^{r}[A_k]\left[\Delta D_{fik}\right] + \sum_{j=1}^{s}[\Delta T_j]\left[D_{wfij}\right]$$
$$+ \sum_{j=1}^{s}\left[T_j\right]\left[\Delta D_{wfij}\right] - \Delta\left(\sum_{m=1}^{t}D_{bm}[\tau_i] + D_{ss}[\tau_i] + \left[\Delta Q_{fi}^{0}\right]\right) \tag{3.15}$$

Under the same assumption as (3.14),

$$\left[\Delta A_k\right] = 0,\ \left[\Delta T_j\right] = 0,\left[\Delta \tau_i\right] = 0$$

Thus, the equation (3.15) becomes,

$$[\Delta Q_{fi}] = \sum_{k=1}^{r}[A_k]\left[\Delta D_{fik}\right] + \sum_{j=1}^{s}\left[T_j\right]\left[\Delta D_{wfij}\right]$$
$$- \sum_{m=1}^{t}\Delta D_{bm}\left[\tau_i\right] + \Delta D_{ss}\left[\tau_i\right] + \Delta\left[\Delta Q_{fi}^{0}\right] \tag{3.16}$$

3.2.1 Definition of thermal disturbance vector
Each item in (3.16) is discussed in detail as followings:

First, the item $\sum_{k=1}^{r}[A_k]\left[\Delta D_{fik}\right]$ can be rearranged as,

$$\sum_{k=1}^{r}[A_k]\left[\Delta D_{fik}\right] = \sum_{k=1}^{r}\Delta D_{f1k}\begin{bmatrix}a_{1k}\\a_{2k}\\\vdots\\a_{ik}\end{bmatrix} + \ldots\sum_{k=1}^{r}\Delta D_{fik}\begin{bmatrix}a_{1k}\\a_{2k}\\\vdots\\a_{ik}\end{bmatrix} \tag{3.17}$$

Where the subscript i stands for the number of heater unit (such as $i = 8$ for a 600MW power unit with eight heaters), and $k = 1 \sim r$, stands for the number of auxiliary steam imposed on the No. i heater unit.

The equation (3.17) is the sum of numbers of items, and each item is the product of a coefficient and a vector. The coefficient is mass quantity of auxiliary flow, which is positive for flowing in and negative for flowing out the thermal system.

The configuration of the vector takes on some well-regulated characteristics, which is exposed in details as followings.

The item $[A_k]$ is a lower triangular matrix, and the vector $\begin{bmatrix}a_{1k} & a_{2k} & \cdots & a_{ik}\end{bmatrix}^{T}$ is shaped by system topology configuration. It consists of the heat exchange quantity in the each heater unit along with the (virtual) path from the access point of the auxiliary steam till hot well, such as the auxiliary D_{f21} in Fig.2, whose access point is the steam extraction pipe of the No.2 heater unit. It hasn't direct heat exchange with the No.1 heater, so the corresponding item $a_{11} = 0$, and along with the dashed (see Fig.2), its heat exchange quantity in No.2 is q_{f21}, and γ_3 in No.3 heater respectively. After the closed heater No.3, it is confluent with condensed water, so the heat exchange quantity is regarded as τ_4. That is, the vector becomes $\begin{bmatrix} 0 & q_{f21} & \gamma_3 & \tau_4 \cdots & a_{i1} \end{bmatrix}^T$. Because the dashed flow path may not be a real flow path of the auxiliary steam considered, it is also called "virtual path" for the convenience to construct the vector for analysis.

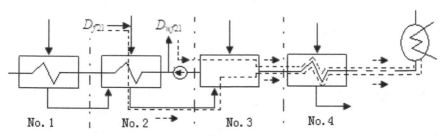

Fig. 2. The schematic illustration for thermal exchange vector of auxiliary steam and water

Second, the item $\sum_{j=1}^{s} \begin{bmatrix} \mathbf{T}_j \end{bmatrix} \begin{bmatrix} \Delta \mathbf{D}_{wfij} \end{bmatrix}$ denotes the thermal disturbance quantity resulted from the auxiliary water flowing in (out) feedwater (condensed) water pipe of all heater units. It is similar with auxiliary steam disturbance except for their flow path. As these auxiliary water are all from or to the water side of the thermal system, their thermal exchange quantity are the enthalpy increase τ_i in each heater unit, such as the auxiliary water D_{wf21} in Fig.2, whose thermal disturbance vector is,

$$\begin{bmatrix} 0 & \Delta \tau_{wf21} & \tau_3 & \tau_4 \cdots & \tau_i \end{bmatrix}^T$$

In this system, $\Delta \tau_{wf21} = \tau_p$,which belongs to the thermal exchange with No.2 heater unit.

Third, $\Delta \begin{bmatrix} \Delta \mathbf{Q}_{fi}^0 \end{bmatrix}$ is the pure heat disturbance, such as pump work, heat transferred by gland heater and so on, whose elements are the direct energy quantity exchanged on the corresponding heater unit.

Finally, the other items in $[\Delta \mathbf{Q}_{fi}]$ are these water flowing in (out) system from boiler side directly, such as continuous blow down, desuperheating spray flow and so on. According to the construction rules for the flow path, their thermal disturbance are regarded as spreading over the entire regenerative heating system. So the thermal exchange of each of them with the heater units is fixed as $\begin{bmatrix} \tau_1 & \tau_2 & \tau_3 & \tau_4 \cdots & \tau_i \end{bmatrix}^T$.

In summary, the vector $\begin{bmatrix} a_{1k}, a_{2k}, \cdots, a_{ik} \end{bmatrix}^T$ is defined as the *thermal disturbance vector*, which is constructed through indentifying the (virtual) flow path of the auxiliary flow considered from its access point till the hot well. Its elements assignment is equal to the heat exchange with each heater unit along with its (virtual) flow path.

3.2.2 Definition of main steam vector

The matrix $[\mathbf{A}]$ is also shaped by system topology configuration as a lower triangular matrix. Inspired by the feature of the *thermal disturbance vector*, it is revealed that the same rule works for the element vector of $[\mathbf{A}]$. For example, in Fig.1 the second extracted steam's access point is the No.2 heater unit. It hasn't direct heat exchange with the No.1 heater, so the first element is 0. Along with the (virtual) flow path from the access point till the hot well, the heat exchange quantity in No.2 is q_2, γ_3 in No.3, and τ_4, τ_5, τ_6 in the No.4, No.5, No.6 heater respectively. The column vector of $[\mathbf{A}]$ is called as the *main system vector* for convenience of equations construction.

3.3 Deduction of system output equations for an thermal power unit
3.3.1 The equation for power output of system
1. The power output equation for the ideal Rankine cycle is:

$$N = D_0(h_0 - h_c)$$

2. For the ideal reheat Rankine cycle:

$$N = D_0(h_0 - h_{eH} + h_r - h_c)$$

3. For the actual cycle of power plant:
Defining $\sigma = h_r - h_{eH}$ gives

$$N = D_0(h_0 + \sigma - h_c)$$

When certain steam $x(D_r, h_r)$ leaves from HP (High Pressure Cylinder), the power output decrease is, $N_x = D_x(h_x + \sigma - h_c)$, and when it leaves from IP(Intermediate Pressure Cylinder) or LP (Low Pressure Cylinder), the power output decrease is $N_x = D_x(h_x - h_c)$. For generalization, the uniform term h_x^{σ} is defined as followings:
When the steam leaves from HP :

$$h_x^{\sigma} = h_x + \sigma - h_c$$

When the steam leaves from IP or LP:

$$h_x^{\sigma} = h_x - h_c$$

For an actual thermal system, the steam leaving from turbine consists of all kinds of extraction steam for heaters and leaking steam from shaft gland. Thus, a complete power output equation for an actual system is obtained according to mass and energy balance.

$$N = D_0(h_0 + \sigma - h_c) - \sum_{i=1}^{p} D_i h_i^{\sigma} - \sum_{n=1}^{u} D_n h_n^{\sigma} + D_{rs} h_{rs}^{\sigma}$$

Written in matrix form:

$$N = D_0(h_0 + \sigma - h_c) - [\mathbf{D}_i]^T[\mathbf{h}_i^\sigma] - [\mathbf{D}_n]^T[\mathbf{h}_n^\sigma] + D_{rs}h_{rs}^\sigma \tag{3.18}$$

Where n stands for number of shaft gland steams, and $h_{rs}^\sigma = h_r - h_c$.

3.3.2 The equation for the heat transferred by boiler

The equation for the heat transferred by boiler for the ideal Rankine cycle is:

$$\dot{Q} = D_0\left(h_0 - h_{fw}\right)$$

For the ideal reheat Rankine cycle:

$$\dot{Q} = D_0\left(h_0 + \sigma - h_{fw}\right)$$

For an actual system, considering all kinds of steam that are not reheated and all kinds of working substance flowing out of and into system from boiler side such as continuous blow down, periodical blow down, the steam for soot blower system, desuperheating spray flow and dereheating spray flow, etc., then the complete equation for the heat transferred by boiler becomes (3.19).

$$\dot{Q} = D_0\left(h_0 + \sigma - h_{fw}\right) - [\mathbf{D}_i]_{c'}^T[\mathbf{\sigma}]_c - [\mathbf{D}_n]_{d'}^T[\mathbf{\sigma}]_{d'} +$$
$$D_{ss}\left(h_{fw} - h_{ss}\right) + D_{rs}\left(h_r - h_{rs}\right) + \sum_{m=1}^{t} D_{bm}\left(h_{bm} - h_{fw}\right) \tag{3.19}$$

3.4 The analytic formula of heat consumption rate

According to (3.10), $[\mathbf{D}_i]$ can be obtained.

$$[\mathbf{D}_i] = [\mathbf{A}]^{-1}\left[D_0[\mathbf{\tau}_i] - [\mathbf{Q}_{fi}]\right]$$

The actual power output can be acquired from site wattmeters, then D_0 and $[\mathbf{D}_i]$ can be obtained from simple iterative calculation using system state equation and equation (3.18).

Substituting $[\mathbf{D}_i]$ into the equation (3.19), then the equation of heat consumption rate $\hat{q}$ can be immediately obtained by $\hat{q} = 3600\dot{Q}/N$.

4. Degradation analysis using state space method

4.1 The state space perspective of thermal system state equation

In our domain, the steady state performance is evaluated for thermal energy system, that is, the dynamic process is neglected and it is focused on the performance evaluation at one steady state. So, equation (3.10) can be regarded as the steady state equation of (2.9). The item $[\mathbf{Q}_{fi}]$ can be regarded as the input vector from auxiliary steam-water flows or the pure heat disturbance. Thus, the next step is to find the output matrix $\mathbf{C}$ and the feedforward matrix $\mathbf{D}$ for system output.

4.2 Deduction of two property vectors

The increment vector of system output can be obtained through equation (3.18) and (3.19).

$$\Delta N = -[\Delta \mathbf{D}_i]^T[\mathbf{h}_i^\sigma] - [\Delta \mathbf{D}_n]^T[\mathbf{h}_n^\sigma] + \Delta D_{rs}h_{rs}^\sigma$$

$$\Delta \dot{Q} = -[\Delta \mathbf{D}_i]_c^T[\mathbf{\sigma}]_{c'} - [\Delta \mathbf{D}_n]_{d'}^T[\mathbf{\sigma}]_{d'} + \Delta D_{ss}\left(h_{fw} - h_{ss}\right)$$

$$+ \Delta D_{rs}\left(h_r - h_{rs}\right) + \sum_{m=1}^{t} \Delta D_{bm}\left(h_{bm} - h_{fw}\right)$$

Then, the matrix equation can be expressed as,

$$\begin{bmatrix} \Delta N \\ \Delta \dot{Q} \end{bmatrix} = -\begin{bmatrix} h_1^\sigma & h_2^\sigma & \cdots & h_i^\sigma \\ \sigma_1^c & \sigma_2^c & \cdots & \sigma_i^c \end{bmatrix}\begin{bmatrix} \Delta D_1 \\ \Delta D_2 \\ \vdots \\ \Delta D_i \end{bmatrix} + \begin{bmatrix} -\Sigma \Delta N_n + \Delta N_{rs} \\ -\Sigma \Delta Q_{d'} + \Delta Q_{rs} + \Delta Q_{ss} + \Delta Q_{bm} \end{bmatrix} \quad (3.20)$$

Where, the change of constant items is zero. $\Sigma \Delta N_n$ and ΔN_{rs} are the change of power output by turbine gland steam and reheater spray water respectively. $\Sigma \Delta Q_{d'}$, ΔQ_{rs}, ΔQ_{ss} and ΔQ_{bm} are the change of boiler heat by HP turbine gland steam, reheater spray water, desuperheating water and boiler blow down respectively.

$$\sigma_i^c = \begin{cases} \sigma & i \le c' \\ 0 & i > c' \end{cases}$$

Replacing the items in the equation (3.20) with the equation (3.14), and defines,

$$[\mathbf{\eta}_i \quad \xi_i]^T = \begin{bmatrix} h_1^\sigma & h_2^\sigma & \cdots & h_i^\sigma \\ \sigma_1^c & \sigma_2^c & \cdots & \sigma_i^c \end{bmatrix}[\mathbf{A}]^{-1} \quad (3.21)$$

Then, equation (3.20) becomes,

$$\begin{bmatrix} \Delta N \\ \Delta \dot{Q} \end{bmatrix} = [\mathbf{\eta}_i \quad \xi_i]^T[\Delta \mathbf{Q}_{fi}] + \begin{bmatrix} \Sigma \Pi_N \\ \Sigma \Pi_Q \end{bmatrix} \quad (3.22)$$

Where $\begin{bmatrix} \Sigma \Pi_N \\ \Sigma \Pi_Q \end{bmatrix} = \begin{bmatrix} -\Sigma \Delta N_n + \Delta N_{rs} \\ -\Sigma \Delta Q_{d'} + \Delta Q_{rs} + \Delta Q_{ss} + \Delta Q_{bm} \end{bmatrix}.$

The matrix $[\mathbf{\eta}_i \quad \xi_i]^T$ just is the transfer matrix of auxiliary thermal disturbance $[\Delta \mathbf{Q}_{fi}]$. It is a constant matrix under a certain steady state of the system. The another item of equation

(3.22) can be regarded as feedforward matrix, which reflects those auxiliary steam or water imposing directly on the system output, such as HP turbine gland steam, spray water.

4.3 The advantages of transfer matrix and thermal disturbance vector for system analysis

So far, we have deduced the transfer matrix $[\eta_i \quad \xi_i]^T$, the main system vector, the thermal disturbance vector and other vectors. The new analysis method can be called vector-based method. The advantages for analyzing a thermal energy system are as followings:

Firstly, it unifies all the analysis formulas for all kinds of auxiliary steam (water) disturbance or pure heat disturbance. That is, we need not to memorize the different formula for different disturbance or to try to discern all its possible results imposed on the system, which may be only competent for an experienced engineer merely.

Secondly, the analysis process can be greatly simplified in terms of different auxiliary flow concerned. For instance, if the whole system is analyzed, that is, the full path of a certain auxiliary steam or water is considered from the source to the destination, so the forward matrix item in equation (3.21) should be considered enough. Otherwise, if only the regenerative heating system is focused on, the necessary calculation is just the product of a constant vector and the thermal disturbance vector of the corresponding auxiliary steam (water) or pure heat disturbance, which can be constructed easily from equation (3.16) and Fig 2.

Thirdly, for certain device performance degradation, only a few local properties are changed and the linear assumption is still satisfying. Thus, equation (3.22) can also be used. such as, terminal temperature difference of heater. Their equivalent thermal disturbance $[\Delta Q_{fi}]$ can be obtained through local balance calculation and equations rearrangement.

5. Case analysis

In order to demonstrate the availability of state space method for the performance analysis of a thermal energy system, the example for an actual 600MW coal-fired power unit is presented. Its system diagram is showed with Fig.3.

Heater No.	h_i (kJ/kg)	tw_i (kJ/kg)	ts_i (kJ/kg)
1	3126.6	1201.1	1077.3
2	3013.9	1052.6	880.8
3	3325.5	863.51	764.4
4	3147.0	725.9	
5	2948.2	569.3	458.1
6	2761.9	435.8	373.4
7	2640.8	351.3	277.7
8	2510.0	255.9	
other Properties (kJ/kg)	h_0=3394.1; h_r=3536.4; h_c=2340.4; h_{cw} =140.7; τ_b=25.3; h_{drum}=2428.73		

Note: h_i is the enthalpy of th ith extracted steam; tw_i is the enthalpy of output water of th ith heater; ts_i is the enthalpy of the drain water of of the ith heater; h_0 is the enthalpy of the main steam; h_r is the enthalpy of the reheated steam; h_c is the enthalpy of the exhausted steam of turbine h_{cw} is the enthalpy of the condensed water; τ_b is the enthalpy increase in feedwater pump; h_{drum} is the saturated enthalpy of drum water

Table 2. System properties under rated load

5.1 The demonstration of calculation process

There are six key steps to accomplish a certain disturbance analysis as followings:

1. According to the construction rule of the *main system vector*, the matrix [A] can be constructed with the parameters given in Tab.2.

$$[A]=\begin{bmatrix} 2049.3 & 0 & 0 & 0 & 0 & 0 & 0 & 0 \\ 196.5 & 2133.1 & 0 & 0 & 0 & 0 & 0 & 0 \\ 116.4 & 116.4 & 2561.1 & 0 & 0 & 0 & 0 & 0 \\ 195.1 & 195.1 & 195.1 & 2577.7 & 0 & 0 & 0 & 0 \\ 133.5 & 133.5 & 133.5 & 133.5 & 2490.1 & 0 & 0 & 0 \\ 84.5 & 84.5 & 84.5 & 84.5 & 84.7 & 2388.5 & 0 & 0 \\ 95.4 & 95.4 & 95.4 & 95.4 & 95.7 & 95.7 & 2363.1 & 0 \\ 115.2 & 115.2 & 115.2 & 115.2 & 137.0 & 137.0 & 137.0 & 2369.3 \end{bmatrix}$$

2. Referring to (3.18), (3.20) to construct the vector $\left[h_1^\sigma, h_1^\sigma, \cdots, h_i^\sigma\right]$ and $\left[\sigma_1^c, \sigma_2^c, \cdots, \sigma_i^c\right]$.

$$\left[h_1^\sigma, h_2^\sigma \cdots h_i^\sigma\right]=[1308.7 \quad 1196.0 \quad 985.1 \quad 806.6 \quad 607.8 \quad 421.5 \quad 300.4 \quad 169.6]$$

$$\left[\sigma_1^c, \sigma_2^c, \cdots, \sigma_i^c\right]=[522.5 \quad 522.5 \quad 0 \quad 0 \quad 0 \quad 0 \quad 0 \quad 0]$$

3. The transfer matrix $\left[\eta_i \quad \xi_i\right]^T$ can be obtained from equation (3.21) with the parameters in Tab. 2.

$$[\eta_i]=\begin{bmatrix} 0.5139 \\ 0.4856 \\ 0.3374 \\ 0.2878 \\ 0.2297 \\ 0.1674 \\ 0.1230 \\ 0.0716 \end{bmatrix} \qquad [\xi_i]=\begin{bmatrix} 0.2315 \\ 0.2449 \\ 0 \\ 0 \\ 0 \\ 0 \\ 0 \\ 0 \end{bmatrix}$$

4. The disturbance vectors can be formed from the equation (3.16) according to its (virtual) flow path, such as the auxiliary steam L, where there isn't heat exchange with heater No.1, No.2 and No.3, the heat exchange with the No.4 heater is $(h_L - tw_5)$ and then become confluent with condensed water through a virtual path to condenser. Thus, for unit quantity of flow, its thermal disturbance becomes,
    $[\Delta q_{fi}]_L =[0,0,0,2440.5,133.5,84.5,95.4,115.2]^T.$

5. Calculating the feedforward item $\begin{bmatrix} \Sigma\Pi_N \\ \Sigma\Pi_Q \end{bmatrix}$ through finding out the source and destination of the auxiliary steam L.

6. Calculating the system output increment that caused by the auxiliary steam L with equation (3.22).

The analysis results for almost all kind of auxiliary flows can be found in Tab.3.

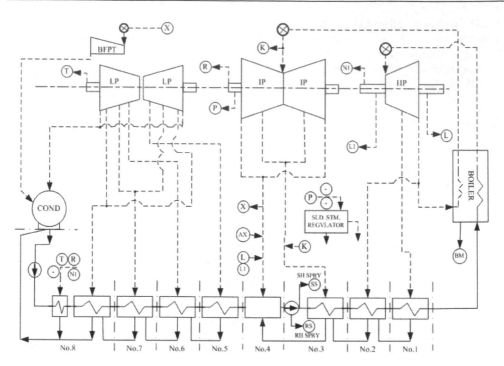

Fig. 3. The thermal system diagram of a 600MW power unit

Aux. No.	enthalpy (kJ/kg)	Quantity Increment² (t/h)	Thermal disturbance vector $[\Delta \mathbf{q}_{fi}]$	Power increase by THR (kW)	Power increase by $\sum \Pi_N$ (kW)	Total increment of power ΔN (kW)	Total increment of boiler heat $\Delta \dot{Q}$ (kJ)
L	3009.8	0.331	$[0,0,0,2440.5,\tau_5,\tau_6,\tau_7,\tau_8]'$	70.532	-109.5886	-39.056	-48.04
L1	3317.3	0.292	$[0,0,0,2748.0,\tau_5,\tau_6,\tau_7,\tau_8]'$	69.399	-121.618	-52.219	-42.38
N1	3317.3	0.013	$[0,0,0,0,0,0,0,3176.6]'$	0.8211	-5.4145	-4.5934	-1.89
SS	751.2	10.0	$[0,0,25.3,\tau_4,\tau_5,\tau_6,\tau_7,\tau_8]'$ $[\tau_1,\tau_2,\tau_3,\tau_4,\tau_5,\tau_6,\tau_7,\tau_8]'$	572.285	0	572.285	1473.9
RS¹	738.55	40.0	$[0,0,12.65,\tau_4,\tau_5,\tau_6,\tau_7,\tau_8]'$	-1268.1	13289.0	12021.0	31087.0
K	3536.4	0.742	$[0,0,2772,\gamma_4,\tau_5,\tau_6,\tau_7,\tau_8]'$	217.707	-246.509	-28.802	0
R	3147.0	0.02	$[0,0,0,0,0,0,3006.3]'$	1.1955	-4.4811	-3.2856	0
P	3147.0	0.114	$[0,0,0,0,0,0,0,0]'$	0	-25.5423	-25.5423	0
X	3147.0	6.495	$[0,0,0,2557.7,\tau_5,\tau_6,\tau_7,\tau_8]'$	-1455.2	0	-1455.2	0
AX	2176.2	10.0	$[0,0,0,1606.9,\tau_5,\tau_6,\tau_7,\tau_8]'$	1464.5	0	1464.5	0
T	2340.4	0.104	$[0,0,0,0,0,0,2199.7]'$	4.5488	0	4.5488	0
BM	2428.73	5.0	$[\tau_1,\tau_2,\tau_3,\tau_4,\tau_5,\tau_6,\tau_7,\tau_8]'$	-450.583	0	-450.583	1817.1

Note: 1. The reheater spray water has an analogy to a low pressure parameters cycle attached to the main steam cycle. Its equivalent efficiency is only 38.67% from the calculation above.
2. The quantity increment of auxiliary flow is ten percent of its rated value except for SS, RS, AX and BM.

Table 3. The results of whole system analysis with kinds of auxiliary thermal disturbance

5.2 Discussion on the calculation results

The fourth column in the Tab.3 consists of the thermal disturbance vector of each auxiliary flow from the definition in the section 3.2.1. For example, there are two thermal disturbance vectors for the desuperheating water, because it imposes two thermal disturbances on the thermal system, one is flowing out the feedwater pipe and another is flowing in boiler. The auxiliary steam P doesn't directly flow into the thermal system, so its thermal disturbance vector becomes zero vector from the rules given in the section 3.2.1.

The fifth column is the power increment obtained from the first item on the left of the equation (3.22), where the flow path from condenser's hot well through regenerative system and boiler till the inlet of turbine is named as thermal heating route (THR), that is, every thermal disturbance vector arises from the THR.

For single auxiliary flow, the matrix equation becomes the product of two vectors plus some simple items, that is, $\Delta N = [\mathbf{\eta}_i]^T [\Delta \mathbf{q}_{fi}] + \Sigma \Pi_N$ and $\Delta \dot{Q} = [\mathbf{\xi}_i]^T [\Delta \mathbf{q}_{fi}] + \Sigma \Pi_Q$. The calculation for every auxiliary flow obeys the rules, which greatly alleviate the difficulties for engineers to memorize many specific formulas.

The specific performance deviation caused by each auxiliary flow deviation can be obtained directly from the total increment of power (ΔN) and the total increment of boiler heat ($\Delta \dot{Q}$). The total deviation comes from the mathematical summation of them. They are identical with those calculated by traditional whole system thermal balance calculation.

6. Conclusion

Process performance monitoring is an overall effort to measure, sustain, and improve the plant and/or unit thermal efficiency, maintenance planning, etc. The decision to implement a performance-monitoring program should be based on plant and fleet requirements and available resources. This includes the instrumentation, the data collection, and the required analysis and interpretation techniques, etc. [14]. This chapter starts from the point of view of system analysis and interdisciplinary methods are employed. Based on vector method of linear system, the new performance monitoring and analysis methodology are proposed, which is intended to achieve an online monitoring/ analysis means beyond traditional periodic test and individual component balance calculation. The main features are as followings:

1. According to error propagation rules and the characteristics of the objective thermo-system, a larger measurements set is adopted to avoid using the primary flow as a necessary input for process performance evaluation, which is impossible to get an accurate calibration for continuous monitoring demand. So good ongoing maintenance of these existing plant instruments and some supplements for traditionally neglected flows measurements may greatly contribute to the achievement of the new methodology, i.e. process performance monitoring is also a kind of management.

2. The system state equations reveal the relationship between the topological structure of thermo-system of a power plant and its corresponding mathematic configuration of steam-water distribution equation, which are deduced for a steady-state system beyond simple mass and energy balance equations, though they are all from the First Law of Thermodynamics

3. The analytic formula of heat consumption rate for thermal power plant indicates that the current heat consumption rate of system can be determined by the system

structure and its thermodynamic properties, as well as all kinds of small auxiliary steam/water flows. It can be easily programmed with the date from existing plant instruments.

4. Based on the research into the assumption for system analysis, the idea of state space model analysis is imported from control theory. The two important vectors, that is, transfer matrix is worked out, which reflects the characteristic of the system itself and hold constant under a steady state condition. Therefore, the system outputs increment is regarded as the results of the disturbance input imposed on the system state space model.

5. The thermal disturbance vector, main system vector, etc. are the new practical approaches proposed here. The regulations to construct these vectors are very comprehensible and convenient, which closely refer to system topology structure and greatly simplify the system analysis process against the traditional whole system balance calculation.

6. The transfer matrix is no longer suitable for the calculation under the circumstances with large deviation of system state or devices degradation, where the current system properties have to be reconfirmed. However, the system state equation holds for any steady state. The vector based method works well on a new reference condition.

In a word, the linearization technique is still an indispensable method for the analysis of such a complex system.

Nomenclature

D mass flow rate, kg/s

h specific enthalpy of the working substance, kJ/kg

N_p total enthalpy increase by feedwater pump shaft work, kJ/kg

N power output, kW

q heat transferred by extracted steam per unit mass in heater. For the closed feedwater heater unit, it is equal to the specific enthalpy difference of inlet steam and outlet drain water. For the open feedwater heater unit, it is equal to specific enthalpy difference of inlet steam and the heater unit's inlet feedwater, kJ/kg

Q rate of heat transfer, kW

$\dot{Q}$ rate of heat transfer by boiler, kW

$[Q_{fi}]$ equivalent vector consisting of the thermal exchange of all auxiliary steam (water) flow or external heat imposed on the main thermal system

$\hat{q}$ heat consumption rate, kJ/kW.h

γ heat transferred by drain water per unit mass in heater. For the closed feedwater heater unit, it is equal to the specific enthalpy difference of the upper drain water and its own drain water. For the open feedwater heater unit, it is equal to the specific enthalpy difference of the upper drain water and the heater unit's inlet feedwater, kJ/kg

τ specific enthalpy increase of feedwater through the heater unit, kJ/kg

τ_p specific enthalpy increase of feedwater through pump, kJ/kg

σ specific enthalpy increase of reheated steam, kJ/kg

Subscripts

0   primary steam

b boiler blowdown

c last stage exhaust steam.

$c^{'}$ total number of the extracted steam leaving before reheating.

d drain water of heater

$d^{'}$ total number of shaft gland steams of HP.

df auxiliary water flow out of or into drain pipe

eH exhaust steam of HP

f auxiliary steam flow out of or into extraction pipe or

fw feedwater

i series number for heater unit

l leakage of steam or water

r reheated steam

rs dereheating spray water

ss desuperheating spray water

sg shaft gland steam

w feedwater or condensate water

wf auxiliary water flow into or out of feed water pipeline

wc condensate water from hot well

x all kinds of steam leaving from turbine

7. References

[1] Performance Test Codes 6 on Steam Turbine[S]. New York: ASME, 2004.

[2] Valero A, Correas L, Serra L. On-line thermoeconomic diagnosis of thermal power plants. In: Bejan A, Mamut E, editors. NATO ASI on thermodynamics and optimization of complex energy systems. New York: Kluwer Academic Publishers; 1999. p. 117–136.

[3] Zhang, C., et al., Thermoeconomic diagnosis of a coal fired power plant. Energy Conversion and Management, 2007. 48(2): p. 405-419.

[4] Zaleta-Aguilar, A., et al., Concept on thermoeconomic evaluation of steam turbines. Applied Thermal Engineering, 2007. 27(2-3): p. 457-466.

[5] Zaleta, A., et al., Concepts on dynamic reference state, acceptable performance tests, and the equalized reconciliation method as a strategy for a reliable on-line thermoeconomic monitoring and diagnosis. Energy, 2007. 32(4): p. 499-507.

[6] Zhang, C., et al., Exergy cost analysis of a coal fired power plant based on structural theory of thermoeconomics. Energy Conversion and Management, 2006. 47(7-8): p. 817-843.

[7] Valero, A., Exergy accounting: Capabilities and drawbacks. Energy, 2006. 31(1): p. 164-180.

[8] Verda, V., L. Serra, and A. Valero, The effects of the control system on the thermoeconomic diagnosis of a power plant. Energy, 2004. 29(3): p. 331-359.

[9] G. Prasad, et al., "A novel performance monitoring strategy for economical thermal power plant operation", IEEE Transactions on Energy Conversion, vol. 14, pp. 802~809, 1999.

[10] K. Nabeshima, T. Suzudo, S., et al. On-line neuro-expert monitoring system for borssele nuclear power plant .Progress in Nuclear Energy, Vol. 43, No. 1-4, pp. 397-404, 2003

[11] Diez, L.I., et al., Combustion and heat transfer monitoring in large utility boilers. International Journal of Thermal Sciences, 2001. 40(5): p. 489-496.

[12] Richard C. Dorf, Robert H B. Modern Control System, Addison Wesley Longman, Inc. 1998.

[13] Yunus A. Çengel, Michael A Boles. Thermodynamics: An engineering Approach (4th Edition). New York: McGraw-Hill 2002

[14] Performance Test Codes PM-2010 [S]. New York: ASME, 2010.

Solar Aided Power Generation: Generating "Green" Power from Conventional Fossil Fuelled Power Stations

Eric Hu[1], Yongping Yang[2] and Akira Nishimura[3]
[1]School of Mechanical Engineering, the University of Adelaide,
[2]North China Electric Power University, Beijing,
[3]Division of Mechanical Engineering, Mie University, Tsu,
[1]Australia
[2]China
[3]Japan

1. Introduction

Nowadays, most power is, and will continue to be, generated by consumption of fossil fuels (mainly coal and gas) which has serious negative impacts on our environment. As a clean, free, and non-depleting source, solar energy is getting more and more attention. However, owing to its relatively low intensity, the application of solar energy for power generation purpose is costly, and the efficiencies of the solar thermal power systems having been developed in which solar energy is used as the main heat source are not satisfactory. In addition, solar energy utilisation is subject to the change of seasons and weather. All of these impede the solar energy's application. How to use solar energy to generate power steadily and efficiently is a problem that needs to be addressed.

In this chapter a new idea, i.e. Solar aided power generation (SAPG) is proposed. The new solar aided concept for the conventional coal fired power stations, ie. integrating solar (thermal) energy into conventional power station cycles has the potential to make the conventional coal fired power station be able to generate green electricity. The solar aided power concept actually uses the strong points of the two mature technologies (traditional Rankine generation cycle with relatively higher efficiency and solar heating at relatively low temperature range). The efficiencies (the fist law efficiency and the second law efficiency) of the solar aided power generation are higher than that of either solar thermal power systems or the conventional fuel fired power cycles.

2. Rankin thermal power generation cycles

Thermodynamically, at a given temperature difference, the most efficient cycle to convert thermal energy into mechanical or electrical energy is the Carnot cycle that consists two isothermal processes (ie. processes 2→3 and 4→1) and two isentropic processes ie. 1→2 and 3→4), as shown in Fig. 1. However, almost all coal or gas fired power stations in the world are operated on so called Rankine as the Carnot cycle is hard to achieve in practice. The basic Rankine cycle, , using steam as working fluid, which is shown in Fig 2, is a modification from the Carnot cycle, by extending the cooling process of the steam to the saturated liquid state, ie. point 3 in Fig 2. In Fig. 2 the process 3→ 4 is a pumping process while the process 4→5→ 1 is the heating process in the boiler. Comparing with Carnot cycle, the Rankine cycle is easier to operate in practice. However, the efficiency of Rankine cycle is lower than that of Carnot cycle. To improve the efficiency of basic Rankine cycle, in real power stations, the Rankine cycle is run as modified Rankine cycles. Three common modifications to the basic Rankine cycle are 1) superheating 2) reheating and 3) regeneration.

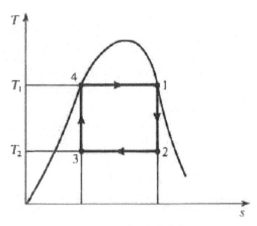

Fig. 1. Carnot cycle for a wet vapour on a T-S diagram[1]

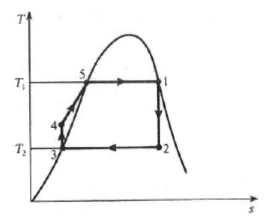

Fig. 2. Basic Rankine cycle using wet steam on a T-s diagram[1]

The regeneration is to extract or called bled off, some steam at the different stages of expansion process, from the turbine, and use it to preheat the feed water entering the boiler. Figure 3 shows a steam plant with one open feed heater, ie. one stage regeneration. In a modern coal or gas fired power station, there are up to 8 stages of extraction and feed water pre-heating existing. Although regeneration can increase the cycle thermal efficiency that is the ratio of power generated to heat input, but the work ratio of the cycle is decreased, which is the ratio of gross work generated to the net power output. In other words, due to the steam extraction or called bled-off, there is less steam mass flow going through the lower stages of the turbine and resulting the power output reduced.

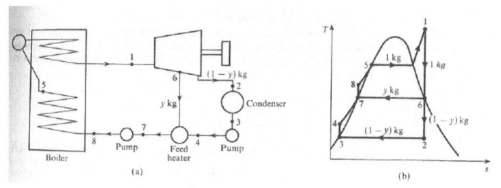

Fig. 3. Steam plant with (a) one stage regeneration and (b) the cycle on a T-s diagram [1]

3. Solar aided power generation

The basis of solar aided power generation (SAPG) technology/concept, is to use solar thermal energy to replace the bled-off steam in regenerative Rankine power cycle. In contrast to other solar boosting or combined power systems, solar energy generated heat (or steam), in SAPG, does not enter the turbine directly to do work. Instead, the thermal energy from the sun is used in place of steam normally extracted from turbine stages for feedwater pre-heating in regenerative Rankine cycles. The otherwise extracted steam is therefore available to generate additional power in the turbine. Therefore the SAPG is capable of assisting fossil -fuelled power stations to increase generating capacity (up to 20% theoretically if all feed heaters are replaced by solar energy) during periods of peak demand with the same consumption of fuel, or to provide the same generating capacity with reduced green house gas emissions.

The SAPG technology is thought to be the most efficient, economic and low risk solar (thermal) technology to generate power as it possesses the following advantages:

- The SAPG technology has higher thermodynamic 1st law and 2nd law ie, exergy efficiencies over the normal coal fired power station and solar alone power station. Preliminary theoretical studies is presented in the following sections.

- Utilizing the existing infrastructure (and existing grid) of conventional power stations, while providing a higher solar to electricity conversion than stand alone solar power stations. Therefore a relatively low implementation cost, and high social, environmental and economic benefits become a reality.

- The SAPG can be applied to not only new built power station but also to modify the existing power station with less or no risk to the operation of the existing power stations.

- The thermal storage system that at present is still technically immature is not necessary. The SAPG system is not expected to operate clock-round and simplicity is another beauty of the SAPG. The pattern of electricity demand shows that nowadays air conditioning demand has a great impact on the electricity load. Afternoon replaces the evening to be the peak loading period in summer. This means that the extra work generated by this SAPG concept is just at the right time. Namely, the solar contribution and power demand are peak at the same time ie. during summer day time.

- The SAPG is flexible in its implement. Depending on the capital a power station has, SAPG can be applied to the power station in stages.
- The SAPG actively involves the existing/traditional power industry into the renewable technology and assist it to generate "green" electricity. It is the authors' belief that without the engagement of existing power industry, any renewable energy (power generating) targets/goals set by governments are difficult or very costly to fulfil.
- Low temperature range solar collectors eg. vacuum tubes and flat plate collectors, can be used in the SAPG. It is a great new market for the solar (collectors) industry.

The benefit of SAPG to a power station can come from either additional power generation with the same fuel consumption ie solar boosting mode, or fuel and emission reduction while keep the same generating capacity ie. fuel saving mode, shown in Fig.4.

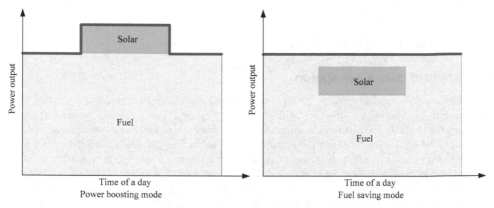

Fig. 4. Two operation modes of solar aided thermal power generation [2]

4. Energy (the first law) advantages of SAPG

In the power boosting operation mode, the thermal efficiency of solar energy in the SAPG system is defined as:

$$\eta_{solar} = \frac{\Delta W_e}{Q_{solar} + \Delta Q_{boiler}} \tag{1}$$

where ΔW_e is the increased power output by saved extraction steam, Q_{solar} is the solar heat input; ΔQ_{boiler} is the change in boiler reheating load, accounting for increases in reheat steam flows. For ΔW_e, Q_{solar} and ΔQ_{boiler}, the unit is kW or MW. In the formula above, no losses (eg. shaft steam loss etc.) have been considered, ie. it is an ideal thermodynamic calculation.

As the SAPG approach is actually makes the solar energy "piggy-backed" to the conventional coal fired power plants, if the power plant itself has a higher efficiency, eg. in a supercritical or ultra-supercritical modern power plant, the solar to power efficiency in the SAPG system can be expected higher.

For example, let's consider three typical temperatures of solar thermal resources at 90°C, 215°C and 260°C. If the solar heat at these 3 temperature levels are utilised to generate power in a solar stand-alone power plant, with SAPG in a typical 200MW subcritical power plant and in a 600MW supercritical (SC) steam plant, respectively, the (the solar to power) efficiencies in these cases are given in Fig. 5. [3]

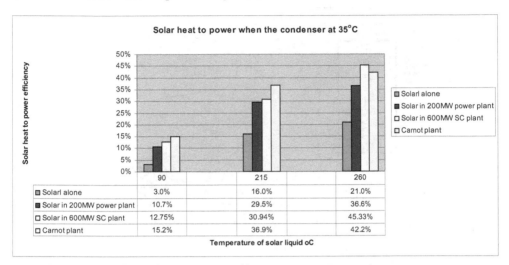

Fig. 5. Comparison of solar heat to power efficiencies in various cycles

In Fig 5, the Carnot efficiency of renewable energy generation is also shown, assuming a heat sink temperature of 35°C. It can be seen that the SAPG approach allows η_{solar} to exceed a Carnot efficiency if the temperature of the solar fluid was the maximum temperature of the Carnot cycle. This demonstrates that η_{solar} is no longer limited by the temperature of the solar fluid, but rather by the maximum temperature of the (power station) cycle.

The advantage of the super-critical power cycle with SAPG is also evident, resulting in a significant increase in efficiency relative to the subcritical cycle, as shown in Fig 5.

5. Exergy (the 2nd law) advantages of the SAPG

There are two ways to evaluate the exergy (the 2nd law) advantages of a SAPG system, ie. net solar exergy efficiency method and Exergy merit index method.[4 and 5]

5.1 Net solar exergy efficiency

To illustrate the exergy advantages of the SAPE, let us examine a single-stage regenerative Rankine cycle with open feedwater heater (Figure 6).

In energy system analysis, not only the quantity, but also the quality of energy should be assessed. The quality of an energy stream depends on the work (or work potential) available from that stream. The capacity for the stream to do work depends on its potential difference with its environment. If a unit of heat flows from a source at a constant temperature T_H to its environment at temperature T_a, with a reversible heat engine, the maximum work the heat energy can do, is called the Availability and also called Exergy of the heat at the temperature T_H. In the case of using solar energy (heat), the exergy in the solar irradiation, Ex_s, is [6]:

$$Ex_s = \left[1 - \frac{4Ta}{3Ts}(1 - 0.28\ln f)\right]Qs \tag{2}$$

where the Ta is the ambient temperature and the Ts is the temperature of the sun, f is the dilution factor which equals 1.3x10⁻⁵, and Qs is the solar heat.

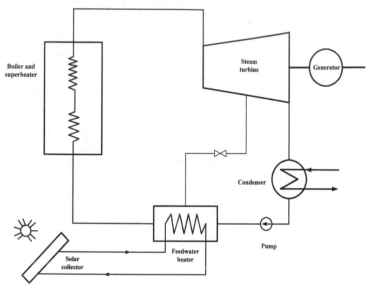

Fig. 6. Single-stage regenerative Rankine cycle with open feedwater heater

In SAPG, the solar heat is used to replace the bled-off steam and heat feed water, so that the solar heat Qs equals:

$$Qs = m \cdot \Delta h = m \cdot c(T_H - T_L) \tag{3}$$

where m is the mass (or flow rate) of the feedwater in the feedwater heater, c is the mean specific heat capacity of the feed water, Δh is the specific enthalpy change of the feed water cross the feedheater.

The net solar exergy efficiency of the SAPG system is then:

$$\eta_{sex} = \frac{\Delta W}{Ex_s} \tag{4}$$

Where ΔW is the extra work generated by the turbine due to the saved bled-off steam.

5.2 Exergy merit index [5]

In the same system shown in Fig 6, the exergy in the extraction steam at T_H is assigned by "e_x", i.e.

$$e_x = w_{max} = q(1 - \frac{T_0}{T_H}) \tag{5}$$

If the temperature of the steam decreases from T_H to T_L, the exergy change of the steam is

$$\Delta e_x = \int_{T_L}^{T_H} (1 - \frac{T_0}{T})\delta q = q - T_0 \int_{T_L}^{T_H} \frac{cdT}{T} = q - T_0 c \ln \frac{T_H}{T_L}$$

$$= c(T_H - T_L) - T_0 c \ln \frac{T_H}{T_L} \tag{6}$$

where c is the mean specific heat capacity of the stream in the temperature range of T_L—T_H. This exergy change of the temperature-changing heat source can also be expressed approximately by a simple form

$$\Delta e_x \approx q(1 - \frac{T_0}{\frac{T_L + T_H}{2}}) \tag{7}$$

To grasp the main points, assume that the steam extracted from the saturated vapour state or from the wet steam region, so the temperature of the extracted steam keeps constant when it transfers heat to the feedwater, while the temperature of the feedwater increases.

If the specific heat capacity of the feedwater c is assumed to be constant (i.e. it is not affected by the temperature's change), then the ratio of exergy increase E_x to heat Q obtained by feedwater (denoted by subscript "w") is:

$$\left(\frac{E_x}{Q}\right)_W = \left(\frac{e_x}{q}\right)_W = \frac{\int_{T_L}^{T_H}(1 - \frac{T_0}{T})\delta q}{h_H - h_L}$$

$$= 1 - \frac{T_0}{T_H - T_L} \ln \frac{T_H}{T_L} \approx (<)1 - \frac{T_0}{\frac{T_H + T_L}{2}} \tag{8}$$

where T_0 is the ambient temperature in K.

The ratio of exergy E_x to heat Q of the extracted steam (at the constant temperature of T_H, denoted by subscript "v") is

$$\left(\frac{E_x}{Q}\right)_V = 1 - \frac{T_0}{T_H} \tag{9}$$

From the heat balance we know that the heat rejected by the extracted steam Q_v equals the heat absorbed by the feedwater Q_w. In addition, the exergy of the extracted steam is very near to the work the steam can do in the turbine. So

$$\frac{\left(\frac{E_x}{Q}\right)_V}{\left(\frac{E_x}{Q}\right)_W} = \frac{Ex_V}{Ex_W} = \frac{W}{Ex_W} \tag{10}$$

From this equation it can be seen that if we supply to the feedwater the same amount of heat with solar energy as the extracted steam did, the saved steam can do work W. If the heat exchange in the heater is reversible, i.e. the heating fluid only releases the same amount of exergy as the feedwater obtained Ex_w, the Equation 10 virtually expresses the ratio of the work we can gain to the exergy cost. In order to assess the merit of using the solar energy in such multi-heat source systems from the view of exergy, we define the available energy efficiency of such scheme as the Exergy Merit Index (*EMI*), which is

$$EMI = \frac{Work_{gain\ from\ steam}}{Exergy_{pay\ by\ solar}} = \frac{W}{Ex_W}$$

$$= \frac{1 - \dfrac{T_0}{T_H}}{1 - \dfrac{T_0}{T_H - T_L}\ln\dfrac{T_H}{T_L}} \approx (>) \frac{1 - \dfrac{T_0}{T_H}}{1 - \dfrac{2T_0}{T_H + T_L}} \tag{11}$$

Since T_L is always less than T_H, the value of *EMI* is always greater than unity. This means that by using the low grade solar energy to replace the high grade extracted steam to heat the feedwater, the work gained from the steam is greater than the available energy given by the solar energy. This is unmatched by any other power systems driven by a single high temperature heat source. Needless to say, this concept is a super energy scheme.

If the solar collector generates vapour, when the temperature of the vapour equals that of the correspondent extracted steam, the *EMI* of the solar energy is unity, which is also unmatched by any other power systems heated by a single high temperature heat source.

Owing to the irreversibility, no matter which way we use to evaluate the exergy advantages of a SAPG system, the benefit will be certainly less than the above values. The study of the following case demonstrates the exergy advantage of the concept in practice.

Here is an example of using the solar energy in a three-stage regenerative Rankine cycle. Assuming that the state of the working fluid at every point of the system does not change with or without solar-aided feedwater heating, only the flow rate changes (with the solar energy aided, the flow rate will increase in the turbine). The pattern is shown in Figure 7. Some important properties are listed in Table 1.

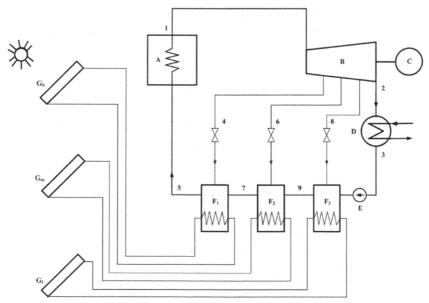

Fig. 7. A three-stage regenerative condensing-steam Rankine cycle

A. boiler and superheater, B. turbine, C. generator, D. condenser, E. pump, F_1, F_2, F_3. feedwater heaters, G_l. low-temperature collector, G_m. medium-temperature collector, G_h. high-temperature collector

Without the solar energy aided, the conventional regenerative Rankine cycle yields work:

$$W_0 = h_1 - h_4 + (1 - m_1)(h_4 - h_6) + (1 - m_1 - m_2)(h_6 - h_8)$$

$$+ (1 - m_1 - m_2 - m_3)(h_8 - h_2) = 1084.96 \quad kJ/(kg \ steam \ in \ boiler)$$

Assuming the ambient temperature is 25°C (298K), and the temperature difference for heat transfer in the condenser is 10°C. When aided heat is used, assuming the average temperature difference for heat transfer in heaters is 10 °C for liquid heat carrier, let us investigate the following cases.

Point in Fig. 7	P (kPa, absolute)	t (°C)	h (kJ/kg)
1. turbine inlet	16500	538	3404.78
2. turbine exhaust	7	38.83	1993.92
3. condensed water	7	38.83	162.7
4. high pressure extracted steam	6000	369.82	3097.15
5. high-stage heater outlet	6000	275.6	1213.4
6. medium pressure extracted steam	1000	179.9	2701.53
7. medium-stage heater outlet	1000	179.9	762.81
8. low pressure extracted steam	101.3	100	2326.44
9. low-stage heater outlet	101.3	100	419.04

Note: the weight fraction of the extracted steam m_1=0.193 m_2=0.1215, m_3=0.0812.

Table 1. Some Properties of the Cycle

From above cases, it can be seen, in Table 2, that using the low temperature thermal energy to heat the feedwater in the regenerative Rankine cycle, the values of exergy efficiency is quite high, comparing to other solar thermal power generation systems.

From the thermodynamic point of view, generally using liquid as the heat carrier for solar energy in these systems is better than using vapour. With SAPG, we can use water (liquid) rather than other low-boiling point substance as working fluid and do not need to use the more sophisticated vapour-generating collectors.

With a little advanced collector, the medium and even high temperature fluid can be made easily. When high temperature heat carrier of the solar energy can be provided, it is suggested to install the multi-stage collectors with different temperature levels to heat the feedwater serially in the multi-stage heaters (see also Figure 7). One advantage of this multi-stage design is the system can be made more flexible so particular stage(s) of extracted steam can be closed according to the load demand in practice. If the vapour/steam can be generated by the (solar) collectors, the pattern of multi-stage collectors with different temperature levels is preferable as the solar net exergy efficiencies of the multi-stage systems are much higher than that of the one-stage system, and the more the stages, the higher the efficiencies.

	Case 1	Case 2
Phase of the heat carrier of the aided energy	liquid	liquid
Highest temperature of the aided energy, °C	110	286
The stage(s) closed	stage 3 only	all stages
Extra work done by the saved steam ΔW, kJ/(kg steam generated in boiler)	27	325.9
The aided solar heat input Qs, kJ/(kg steam generated in boiler)	175.72	1050.7
Thermal efficiency of the solar energy in the aided system η_I ($\eta_I=\Delta W/Q$), %	15.37	30.12
Exergy contained in the aided solar energy Ex_s, kJ/(kg steam generated in boiler), using eq (1):	163.1	975.4
Net solar exergy efficiency in the SAPG, %, η_{sex} in eq (3)	16.6	33.4
Exergy cost (payed by the aided solar energy) E_x, kJ/(kg steam generated in boiler)	26.746	321.4
EMI of the solar energy in the aided system ($EMI=\Delta W/E_x$), %	101	101.4
Work increased (Comparing with the conventional regenerative Rankine cycle), ($\Delta W/W_0$), %	2.5	30.04

Table 2. Analyses on solar aided systems

6. Solar percentage

In the SAPG case, the relative contribution of solar energy to the total power output from the plant is shown in Fig. 8 when the solar energy is assumed to be available around the clock with storage. These calculations show that for a solar-fluid at 215 °C, which is not very high and relatively easy to achieve, the solar contribution to total power is about 16%.

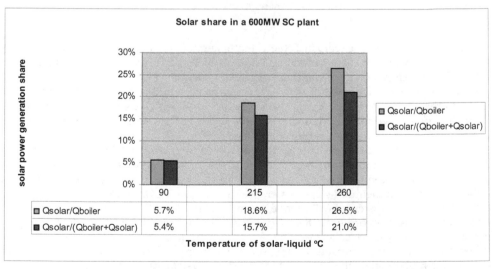

Fig. 8. Solar thermal share of power generation in a RAPG 600MW supercritical plant

Increasing the solar input temperature to 260 °C increases the contribution of solar energy to about 21%. For the reduced solar input temperature of 90 °C (from a non-concentration solar collector), the SAPG case provides a much lower total contribution to power, at about 5%. For the solar-thermal case, a time fraction needs to be considered when calculating the solar contribution to total power out put, because the availability of solar depends on seasons and locations.

7. A real case study [7]

The solar heat at various levels can be used in this SAPG system. The high temperature solar heat from the parabolic trough solar collector at nearly 400°C can be easily used in the first couple of stages of high pressure feedwater heaters, while the low temperature solar heat from flat vacuum collectors at 200°C or less can be used in the lower stages of low pressure feedwater heats. The solar heat can be used in either closed or open (deaerator) types of feedwater heaters and can replace the extraction steam either fully or partly in a particular stage. Therefore, the SAPG technology is sometime also called multi-points and multi-levels solar integration.

A typical regenerative and reheating Rankine steam system has been shown in Fig. 9, which is N200-16.8/530/530 system made by Beijing Beizhong turbine factory in China. The boiler is composed of furnace, drum, risers, superheaters, feedwater heaters and economizer. The combustion of coal takes place in the boiler. The unsaturated feed-water from condenser enters the boiler after going through four low-pressure feedwater heaters (3-6 in Fig.9), two

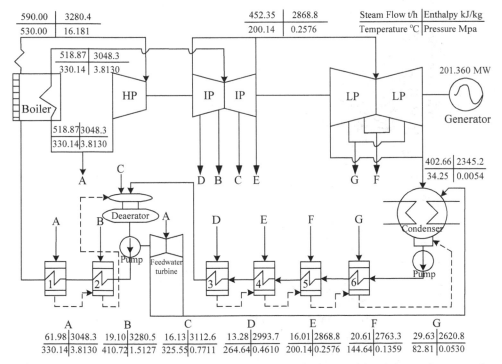

Fig. 9. Schematic diagrams and thermal balance of a 200 MW coal-fired thermal power plant

high pressure feedwater heaters (1 and 2 in Fig.9) and a deaerator. The superheated steam from the boiler enters the high pressure turbine to generate power. After reheated in the boiler, the steam expands further through intermediate pressure and lower pressure stages of the turbine. In the end, the final exhaust steam is condensed in the condenser. The deaerator is actually an open type feedwater heater to preheat the feedwater and remove the oxygen. The feedwater heaters are closed type heaters. The aim of extracting steam from turbine to preheat feed water is to increase overall thermal efficiency of the system.

As stated before, the flexibility is one of advantages the SAPG has. In the SAPG, the solar replacement of extraction steam in a particular feedheater does not need to be 100%, instead it can be any percentage from 0% to 100%.

In terms of working fluids (in solar collector) selection, there are two options, one is using boiler quality of water/steam directly and the other is using something like thermal oil. The additional heat exchangers need to be installed in parallel with the feedwater heaters in the later option. However, in terms of the energy analysis later in the paper, the two options have no fundamental differences.

A mathematical simulation model has been developed to carry out the case study, the results are shown in Figures 10-12 below[7]:

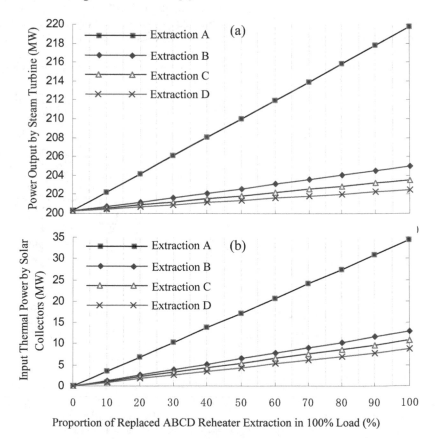

Fig. 10. The output characters vs. solar percentages at A, B, C and D in 100% load

It can be seen from Fig.10(a) that the increased power output is nearly 20 MW, i.e. the total plant power output reaches to 220 MW when the extracted steam (of 62t/h) from A is completely replaced by solar heat. At the same time, the Fig.10(a) shows the additional power generated will be less if the replacement is at locations B, C or D, as expected, because the quality i.e. temperature of solar input required is lower than steam of extraction A. Fig.10(b) shows the solar heat demands for the cases in Fig.10(a). Certainly, more solar input will replace more extracted steam and generate more additional power. For example, when replacing A and D extractions completely, the increased power outputs are 19.72 MW and 2.55 MW, respectively. The ratio is 7.7, much greater than the solar energy input ratio of 3.9. It is concluded that it is more efficient to replace the higher stage of bled-off steam, if the solar heat temperature is able to do so.

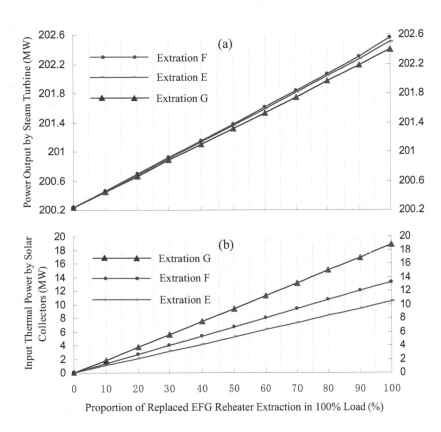

Fig. 11. The output characters vs. solar percentages at E, F and G in 100% load

The extracted steams from locations E, F and G are classified as the low temperature group, the extraction steam temperatures at these locations (in this case) are 200.14°C, 144.64°C and 82.81°C, respectively.

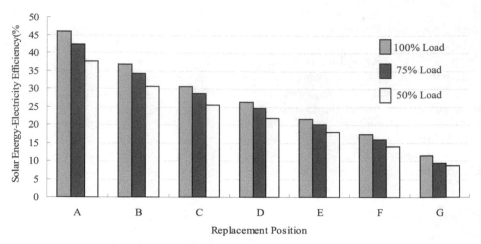

Fig. 12. Solar heat to electricity efficiency in different loads

The solar to power efficiencies in 100% replacement at various generating capacities, as shown in Fig.12, are calculated using Eq. 1. It shows the solar efficiency in the SAPG cases is much higher that in the other solar thermal power generation systems using the same quality/temperature of solar heat.[3,5] The maximal efficiency (45%) occurs at location A where the solar heat temperature is just about 330°C, when the plant operats at full capacity. At the low temperature sections i.e. location G where the solar heat input can be lower than 100°C, the solar efficiency can still be at about 11.5% .

8. Discussions

8.1 Turbine working under off design condition

When the extract steam is replace by solar fluid in SAPG system, the steam mass flows through the lower stage turbines are changed. In other word, the lower stage turbines are actually working at off design conditions. Under this condition, the Stodola's law (Ellipse law) is often used to estimate the pressure changes (due to mass flow change) in the lower stages turbines. The Stodola's law can be written in the form below[8]:

$$\frac{D_1}{D_{10}} = \sqrt{\frac{p_1^2 - p_2^2}{p_{10}^2 - p_{20}^2}} \sqrt{\frac{T_{10}}{T_1}}$$

Where the D_1 is the design flow rate, the D_{01} is the off-design flow rate, p_1 is the pressure inlet pressure at design condition, p_2 is the pressure of outlet pressure at off-design condition, p_{10} is the inlet pressure at design condition, p_{20} is the outlet pressure at off-design condition. T_1 and T_{10} is the inlet temperature at design condition and off-design condition.

However, when estimating benefits of SAPG plant, considering the turbine working at off-design condition or not would not have too much impact. The difference is less than 1% [9]. Therefore the results in the previous sections of this chapter and the literatures did not consider the turbines working under the off design conditions.

8.2 Limits of steam mass flow changes in turbines

For a conventional 200MWpower plant, normally the maximum capacity of turbine is nearly 220MW. If SAPG is used in such a plant, the pant is run at its near maximum capacity, which may impair the safety for the plant.

However, according to the recent statistics (in China), the majority of power stations have retrofitted the trough-flow structure of the turbine to increase the rated capacity. For example, 200 MW coal-fired power plants are retrofitted to 220MW and maximum generating capacity is then increased to nearly 235 MW or more. Therefore, SAPG, ie. the replacement of the extracted steam, can be realized in the retrofitted plants more easily.

9. Conclusions

The advantages of Solar Aided Power Generation concept in the aspects of its energy and exergy, have been shown in this chapter. By using solar energy to replace the extracted steam in order to pre-heat the feedwater in a regenerative Rankine plant cycle, the energy and exergy efficiencies of the power station can be improved. The higher the temperature aided heat source is, the more beneficial the system can generate. It can be seen that the low-grade energy, eg. solar heat from non-concentrated collectors (and other possible waste heat), is a valuable source of work if it can be used properly. This "aided" concept is different from other solar boosting and hybrid power generation concepts as the solar heat in the form of hot fluids (oil or steam) does not enter the steam turbine directly, thus the solar heat to power conversion efficiency would not limited by the temperature of the solar fluid.

The SAPG has special meanings for solar energy. For in summer weather, both the solar radiation and the electrical load demand peak, and it is easy to make heat carrier in different temperatures with different type of collectors. So the increased solar radiation can supply the increased energy to meet the increased power demand. In addition, the solar aided system can also eliminate the variability in power output even without thermal storage system. The concept of the solar aided power system is really a superior energy system and is a new approach for solar energy power generation.

10. References

[1] Eastop & McConkey, Applied Thermodynamics for Engineering Technologists, Longman Scientific & Technical, 5th Edition, 1993.

[2] Yongping Yang, Qin Yan, Rongrong Zhai and Eric Hu, An efficient way to use medium-or-low temperature solar heat for power generation-- integration into conventional power plant, Applied Thermal Engineering 31 (2011), pp. 157-162.

[3] Eric Hu, Graham J. Nathan, David Battye,Guillaume Perignon, and Akira Nishimura, An efficient way to generate power from low to medium temperature solar and geothermal resources, Chemeca 2010, 27-29 Sept. 2010, Adelaide, Australia. Paper #138, ISBN: 978-085-825-9713.

[4] Eric Hu, YP Yang, A Nishimura and F Yilmaz, " Solar Aided Power Generation" Applied Energy, 87 (2010) pp2881-2885.

[5] Y.You and Eric Hu: Thermodynamic advantage of using solar energy in the conventional power station, Applied Thermal Engineering, Vol. 19, No. 11, 1999.

[6] H. Zhai et al, Energy and exergy analyses on a novel hybrid solar heating, cooling and power generation system for remote areas, Applied Energy, Volume 86, Issue 9, 2009, Pages 1395-1404.

[7] Q.Yan, YP Yang; A. Nishimura, A. Kouzani; Eric Hu, "Multi-point and Multi-level solar integration into conventional power plant" Energy and Fuels , 2010, 24(7), 3733-3738.

[8] A.Valero, F.Lerch, L.Serra, J.Royo, Structureal theory and thermodynamic diagnosis, Energy Conversion and Management vol.43 (2002), p1519–1535

[9] JY Qin, Eric Hu and Gus Nathan, "A modified thermodynamic model to estimate the performance of geothermal aided power generation plant", The 2011 International Conference on Energy, Environment and Sustainable Development (EESD 2011), 21-23 Oct. 2011, Shanghai, China.

Part 2

Fuel Combustion Issues

3

Fundamental Experiments of Coal Ignition for Engineering Design of Coal Power Plants

Masayuki Taniguchi

Hitachi Research Laboratory, Hitachi, Ltd.
7-1-1 Omika-cho, Hitachi-shi, Ibaraki-ken,
Japan

1. Introduction

1.1 Background

Pulverized fuel combustion systems are widely used in thermal power plants. Plant performances vary with solid fuel properties, but it is difficult to evaluate the effects of fuel properties using large scale facilities. Fundamental experimental techniques are required to evaluate the effects for actual systems. Small scale experiments are comparatively low-priced and simple, and they require a comparatively short time and small amount of coal. In the present study, we focused on ignition properties. First, we developed a fundamental experimental technique to examine ignition performances of various coals. Then, we examined the common point and the differences of ignition phenomena of the actual systems by fundamental experiments. Finally, some application examples were considered.

Ignition properties are fundamental combustion performance parameters for engineering design of combustion systems. Many fundamental research studies have been carried out regarding coal ignition. Ignition temperature, flammability limit concentration (explosion limit concentration) and burning velocity (flame propagation velocity) are important ignition performance parameters. We examined a technique for getting these fundamental ignition parameters to apply burner design for pulverized fuel combustion systems.

The ignition temperatures of coals have been extensively measured. For example, a thermogravimetric technique [1, 2] and an electrically heated laminar-flow furnace (drop-tube furnace) technique [3] have been used. The ignition temperature is useful to grasp differences in fuel properties and surrounding gas compositions. However, there are some problems for applying the ignition temperature data to actual burner designs. For example, the ignition temperature sometimes decreases when the particle diameter is increased. Such results would lead to the idea that flame stabilization becomes easy when particle diameter is increased. However, decreasing the particle diameter is very important to obtain a stable flame for actual burner systems.

In industries handling powders, the prevention of dust explosions is important for ensuring safety. Measuring the explosion limit concentration is important and some standard experimental devices have been applied to this [4, 5]. Explosion limit concentration has been measured for various coals and experimental conditions. It is technically difficult to raise fuel concentration for a pulverized coal firing boiler, so measurement of the lean explosion limit concentration is important. Generally, lean explosion limit concentration is low when the particle diameter is fine [4]. This tendency accords with the phenomenon experienced in actual systems. Explosion limit concentrations are basic data for burner designs. At the very least, coal concentration should be larger than the lean explosion limit concentration. However, there are two problems for application.

1. The flames formed by explosion limit experiments are unsteady. It is necessary to clarify common points and differences with the continuously formed burner flame.
2. Influence of heat loss on flame stability is large for pulverized coal flames.

For actual boilers, the coal flames are surrounded by the furnace wall. The furnace wall temperature is several hundred degrees Celsius for a water wall, and it is larger than one thousand degrees Celsius for a caster wall. On the other hand, wall temperatures are usually room temperature for the experimental devices used to measure explosion limit. It is necessary to establish a procedure to correct for the influence of heat loss from the flame to surroundings.

Burning velocity (flame propagation velocity) has been used widely for gas combustion as a basic physical quantity [6, 7]. Burning velocity affects the burner design. Generally, velocity of fuel and air injected from the burner are regulated in response to burning velocity. However, there have been only a few studies about these topics for pulverized coal combustion. Flame propagation velocities of pulverized coals were measured in the microgravity condition [8, 9]. Fujita et al. [8] examined the effect of surrounding oxygen concentration and pressure on flame propagation velocity, while Suda et al. [9] examined flame propagation velocities for oxy-fuel combustion conditions. Chen et al. [10] and Taniguchi et al. [11] developed laser ignition equipment to examine flame propagation velocity of pulverized coals. As well, Taniguchi et al. [12] examined the relationship between flame propagation velocity and lean flammability limit for various coals.

In the present study, we introduce procedures to apply these fundamental experiment results to burner design. A basic model of the flame propagation was examined at first. Then, effects of experimental conditions, such as coal properties, particle diameters, and surrounding gas compositions, were examined. The laser ignition experiments provided an unsteady flame. Ignition phenomena obtained with the laser ignition experiments were compared with those of continuous flames. Some correction procedures were introduced to apply the fundamental data to actual burner designs. Finally, a case study was introduced to obtain stable combustion for some biomass fuels.

1.2 Ignition and flame propagation phenomena for pulverized coal combustion

Recently, reduction of CO_2 emissions is required for coal fired thermal power plants. To achieve this, various approaches have been taken. Ultra supercritical power plants have been developed for improvement of power efficiency [13]. Oxy-fuel combustion technology is being pursued for carbon capture and storage [14]. Further reduction of the environmental load, such as NOx reduction, is also still required [15].

Numerical simulations such as computational fluid dynamics (CFD), are very important to design such new technologies. Figure 1 shows an example of a CFD calculation for a pulverized coal firing boiler [16-18]. The temperature distribution is shown in the figure as well as examples of items targeted for the evaluation. Numerical analyses were first applied to evaluate heat absorption by the furnace wall [19], since then they have been applied to such environmental performance factors as NOx emission [20, 21] and to control furnace wall corrosion [22]. Evaluation of flame stability, such as prediction of blow-off limit, is possible, but, this evaluation is relatively difficult [23, 24].

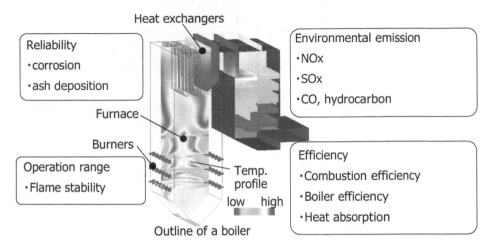

Heat exchangers

Reliability
- corrosion
- ash deposition

Environmental emission
- NOx
- SOx
- CO, hydrocarbon

Furnace

Burners

Operation range
- Flame stability

Temp. profile

low   high

Efficiency
- Combustion efficiency
- Boiler efficiency
- Heat absorption

Outline of a boiler

Fig. 1. Schematic of a pulverized coal fired boiler and results that can be calculated by using CFD.

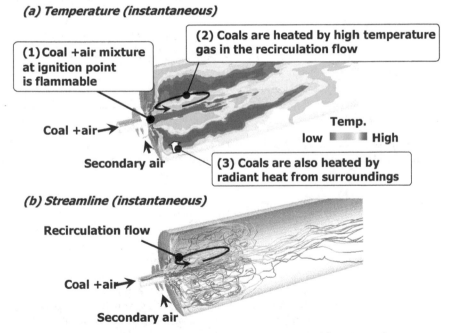

(a) Temperature (instantaneous)

(1) Coal +air mixture at ignition point is flammable

(2) Coals are heated by high temperature gas in the recirculation flow

Coal +air

Secondary air

Temp.

low   High

(3) Coals are also heated by radiant heat from surroundings

(b) Streamline (instantaneous)

Recirculation flow

Coal +air

Secondary air

Fig. 2. Examples of detailed calculated results of pulverized coal combustion.

Figure 2 shows the temperature distribution (a) and streamlines (b) in a burner neighborhood. A recirculation flow is formed around the burner exit. Pulverized coals are ignited by the high temperature gas in the recirculation flow. It is necessary to meet following three conditions to form a stable flame.

1. The coal + air mixture at the ignition point is flammable.
2. Coal particles are heated by high temperature gas in the recirculation flow.
3. Coal particles are also heated by radiant heat from the surroundings.

The conditions (1) and (3) were investigated by fundamental experiments. Flame stability can be examined more precisely by combining the fundamental experiments and CFD calculations.

Table 1 lists examples of coals used for power plants and their properties that were examined. Recently, the range of studied fuel properties has been spreading. The establishment of combustion technology that can support a wide range of fuel properties is required.

Fuel type	Higher calolific value	volatile matter	Fixed carbon	ash	C	H	O	N	S
	(MJ/kg, dry)	(wt%, dry)			(wt%, dry ash-free)				
biomass	19.7	76.4	20.4	3.2	49.8	6.3	43.5	0.4	0.01
Lignite	26.5	36.8	35.2	28.0	70.7	4.9	23.1	1.0	0.3
sub-bituminous	26.9	44.5	41.4	14.1	76.3	6.4	15.3	1.3	0.7
sub-bituminous	27.2	42.5	49.0	8.5	69.1	5.4	23.8	1.1	0.6
hv-bituminous	32.4	37.6	55.1	7.3	83.6	5.5	7.7	1.6	1.5
hv-bituminous	29.7	32.5	53.2	14.3	83.4	5.4	8.8	1.9	0.5
mv-bituminous	30.9	21.8	70.8	7.4	87.6	4.7	4.9	2.1	0.8
lv-bituminous	29.3	14.3	68.8	16.9	89.0	4.3	4.6	1.6	0.4
petroleum coke	36.8	11.8	85.9	2.5	86.3	3.9	1.3	2.9	5.7

Table 1. Examples of coal used for power plants and their properties

Object of this chapter is to develop a model to predict lean flammability limit and flame propagation velocity for pulverized solid fuels, and to apply the model to engineering design for burner systems.

2. Laser ignition experiments

2.1 Experimental equipment

Figure 3 shows a schematic of the laser ignition equipment [10-12, 25]. Uniformly sized pulverized coal particles were suspended in a laminar upward flow and rapidly heated by a single-pulsed YAG laser. Velocity of the upward flow was controlled according to the particle diameter. The heated pulverized coal particles were burned in the quartz test section (50mm cross section area). Emissions from the igniting and burning particles were detected with three photomultiplier tubes (PMTs) and the events were concurrently recorded with a high speed camera.

A He-Ne continuous sheet laser (sheet width 3x10 mm; energy flux around 2×10^{-2} W/m^2) was irradiated horizontally at the ignition point. The particle concentration was measured from the intensity of particle scattering by the He-Ne laser. Another continuous laser (copper vapor laser: maximum laser power 15W; wavelength 488 and 515 nm) was used to reveal the effect of radiant heat loss on the ignition characteristics. The radiant heat loss was

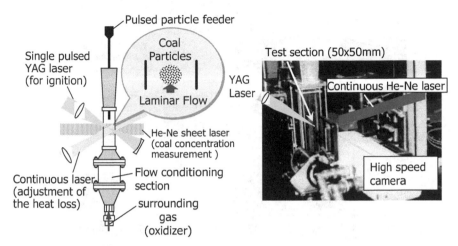

Fig. 3. Schematic drawing of laser ignition experiments

larger for the basic equipment because the quartz wall of the cross-sectional area was kept at room temperature. The effect of radiant heat loss was examined by varying the energy flux of the copper vapor laser. The laser beam was supplied through an optical fiber. The beam diameter was controlled using a collimating lens and concave mirror. Usually, the beam diameter was 15 mm around the ignition point.

The advantage of the laser ignition experiments is the ability to observe ignition phenomena which are similar to actual combustion phenomena, by using a very little amount of coal sample. Both flame propagation velocity and lean flammability limit can be obtained. Observed flame images are shown in Fig.4. Flame propagation velocity was analyzed by increasing the flame radius [11]. Photos of the flame of a hv-bituminous coal (high volatile content) and a lv-bituminous coal (low volatile content) are shown in the figure. The flame of the hv-bituminous coal grew faster than that of the lv-bituminous coal. Flame propagation velocity was large when volatile content was high. Both flames of fine particles and coarse particles are shown for lv-bituminous coal. When particle diameter was fine, flame propagation velocity was large. Lean flammability limit was analyzed by measuring the lower limit of particle concentration at which the flame could grow [11]. When coal concentration was high; almost the same as that for actual systems, the flame moves from the particle directly heated by the pulsed YAG laser to neighboring particles; in this way the flame grew. We defined these phenomena as flame propagation [11, 25]. Only particles directly heated by the pulsed laser burned when the coal concentration was low. Under this condition, we could observe ignition phenomena of a single particle [10, 26].

It is important to have particles stand still as much as possible, in order to obtain reproducible data. Prior to experiments, we observed the stationary state of the particles. Figure 5 shows photos of floating particles near the ignition point when suspended in N_2 flow. Scattering of the He-Ne sheet laser was used for observation. Particles were scattered symmetrically judging from the photos. Particle concentration was evaluated by the intensity of scattering light. Optimum velocity of the upward flow and timing of the laser irradiation were decided, so that variation of particle concentration with time became small just before the YAG laser irradiation.

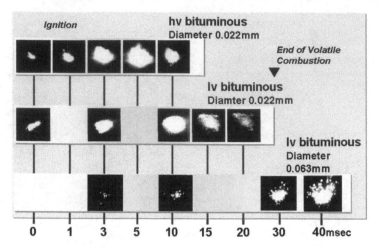

Fig. 4. Examples of flame propagation phenomena obtained from the laser ignition experiments.

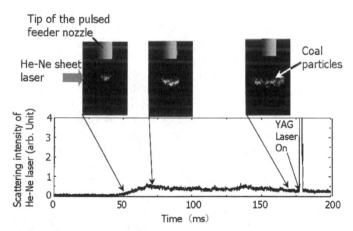

Fig. 5. Observation of floating particles.

2.2 Basic model for ignition and flame propagation

Figure 6 summarizes the basic phenomena of flame propagation. One of two particles burns first, then, the other particle is ignited by the heat of combustion of the one burning particle. When the first particle ignites, volatile matter is pyrolized. A volatile matter flame is formed around the first particle. The flame grows due to volatilization, and the flame heats the next particle which has not ignited yet. Flame propagation is observed if the first burning particle can transfer the flame to the next particle before the volatile matter combustion of the first particle has finished. We defined the distance between particles as d and the time of flame propagation as s. Flame propagation velocity Sb was defined as the value of d divided by s. Relationships between d and s under various experimental conditions were analyzed.

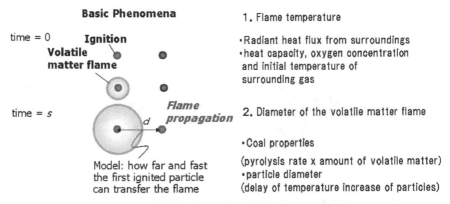

Basic Phenomena

1. Flame temperature

time = 0   **Ignition**
Volatile
matter flame

• Radiant heat flux from surroundings
• heat capacity, oxygen concentration
  and initial temperature of
  surrounding gas

time = *s*   *Flame*
                 propagation

2. Diameter of the volatile matter flame

• Coal properties

Model: how far and fast
the first ignited particle
can transfer the flame

(pyrolysis rate x amount of volatile matter)
• particle diameter
(delay of temperature increase of particles)

Fig. 6. Basic model for flame propagation.

An example relationship between coal concentration and flame propagation velocity is shown in Fig.7 [12]. Coal concentrations and flame propagation velocities are shown as normalized values. The coal concentration was in inverse proportion to the third power of the distance *d*. When the coal concentration increased, flame propagation velocity increased. But there was an upper limit value (*Sb-max*) to the flame propagation velocity. The flame propagation velocities were almost zero at the lean flammability limit. Absolute values of *L* and *Sb-max* were found to vary with coal properties and burning conditions. An example relationship between *L* and *Sb-max* is shown in Fig. 8. Lean flammability limit; *L* was inversely proportional to maximum flame propagation velocity; *Sb-max* [12].

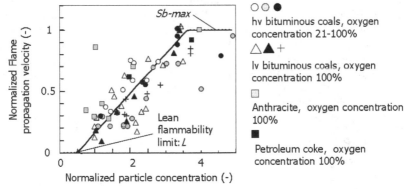

hv bituminous coals, oxygen
concentration 21-100%

lv bituminous coals, oxygen
concentration 100%

Anthracite, oxygen concentration
100%

Petroleum coke, oxygen
concentration 100%

Fig. 7. Relationship between coal concentration and flame propagation velocity [12].

Data on bituminous coals and anthracites of the same particle diameter were obtained under various burning conditions [12], and the effect of particle diameter was examined [11, 25]. Relationships between *L*, *Sb-max* and particle diameter *Dp* are summarized by Eqs. (1)-(3).

$$Sb\text{-}max \propto \ 1\ /\ L \tag{1}$$

$$L\ /\ Sb\text{-}max \propto \ Dp^2 \tag{2}$$

$$L \propto \ Dp^{1.5} \tag{3}$$

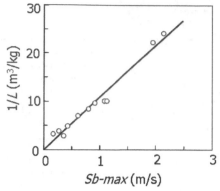

Fig. 8. Relationship between maximum flame propagation velocity Sb-max and lean flammability limit L [12].

2.3 Effects of experimental conditions on flame propagation performances for laser ignition experiments

Ignition properties of solid fuel vary significantly with fuel properties. One of the most important things is to keep the fuel concentration within the flammable condition, to form a stable flame. Effects of fuel properties on lean flammability limit were examined experimentally at first. Figure 9 shows the lean flammability limit L at the same diameter with different volatile contents of coals. The evaluation method of lean flammability limit has been shown elsewhere [11, 25]. The vertical axis of Fig. 9 is the reciprocal of the lean flammability limit concentration. Lean flammability limits were low for high volatile coals.

The growth rate of the volatile flame is usually large for high volatile coals, because the pyrolysis rate is usually large. The flame propagation time s in Fig. 6 was short. The amount of volatile matter was large for high volatile coals, so that the volatile flame which formed around a particle became large. The flame could be transmitted in a short time from one burning particle to another, even though the distance between the particles d was large.

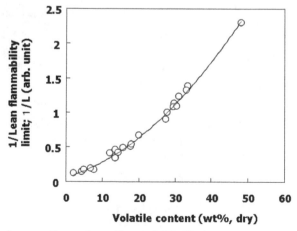

Fig. 9. Effects of coal properties on lean flammability limit.

The lean flammability limit of solid fuels is related to the pyrolysis rates and the amount of pyrolized fuel [24]. The lean flammability limit can also be evaluated by using pyrolysis experiments. Examples of pyrolysis calculations for various fuels are shown in Fig.10. The calculation method has been shown elsewhere [24, 25]. The relationship between heating time and amount of pyrolized volatile matter was calculated when the particle was heated at a rate of 20000K/s. The distributed activation energy model (DAEM) was used for the pyrolysis [24-29]. The pyrolysis rate dV/dt was calculated by Eq. (4).

$$\frac{dV}{dt} = V^\infty \int_0^\infty Av \, e^{-Ev/RTp} \, exp \, [- \int_0^t Av \, e^{-Ev/RTp(t')} dt \,] \, f(Ev) \, dEv \qquad (4)$$

The initial mass of volatile matter V^∞ was evaluated by Flashchain [29]. The function $f(Ev)$ shows the distribution of the activation energies and the frequency factor (Av). As shown in equations (5) and (6), $f(Ev)$ is expressed as a summation with more than one normal distribution function. The function $f(Ev)$ gives results that approximately agree with the experimental results for each coal type.

$$f(Ev) = \sum_{i=1}^n \frac{a_i}{\sqrt{2\pi}E\sigma_i} \, exp \, [-(E-Eav_i/2E\,\sigma_i^2] \qquad (5)$$

$$\sum_{i=1}^n a_i = 1 \qquad (6)$$

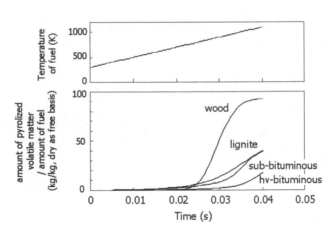

Fig. 10. Examples of pyrolysis calculation results for evaluating lean flammability limit.

In general, pyrolysis started at low temperature and the amount of pyrolized volatile matter increased when volatile content of fuel increased, as shown in Fig. 10. We assumed that L was in inverse proportion to the amount of pyrolized volatile matter

Figure 11 compares estimated L values based on the pyrolysis calculation with measured values. The pyrolysis rate constant was fitted by thermogravimetric analyses [27]. The estimated values agreed with the experimental values. Pyrolysis rates and amount of the pyrolyzed fuels was strongly related to lean flammability limits.

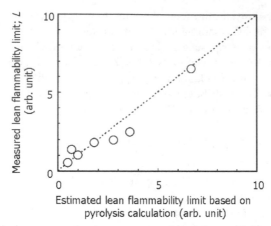

Fig. 11. Relationship between pyrolysis property and lean flammability limit. Estimated flammability limit based pyrolysis calculations are compared with measured lean flammability limits.

The particle diameter also strongly influences lean flammability limit. When the diameter is uniform, the lean flammability limit is proportional to the 1.5th power of the particle diameter and the maximum distance between particles for flame propagation is proportional to the square root of the diameter [11].

The solid fuels are ignited by the following.

1. The particles are heated from the surroundings
2. The particle temperature increases, then, volatile matter is pyrolized.
3. The concentration of pyrolized volatile matter increases, then, pyrolized volatile matter and particles ignite.

When the particle diameter is large, a rise of particle temperature in (2) becomes slow.

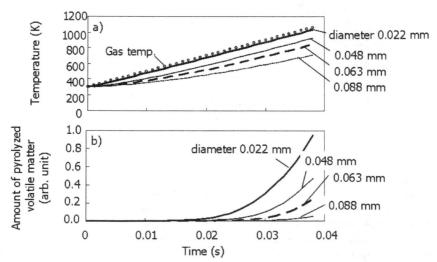

Fig. 12. Effect of particle diameter on pyrolysis property.

Effect of particle diameter on pyrolysis characteristics was examined. The results are shown in Fig.12. Pyrolysis processes were calculated when the surrounding gas temperature was heated at a constant rate. Heating rate of surrounding gas was 20000K/s. Figure 12(a) shows time changes of the amount of pyrolyzed volatile matter, and Fig. 12(b) shows time changes of the particle temperature.

The method of calculation has been shown elsewhere [27]. Heat balance of particles was written as

$$C_p \rho V_p \frac{dT_p}{dt} = Sh(T_g - T_p) - S \varepsilon \sigma(T_p^4 - T_w^4) \tag{7}$$

where C_p is specific heat of particles, h is convective heat transfer coefficient, T_g is gas temperature, T_p is particle temperature, T_w is wall temperature, S is external surface area of particles, ρ is particle density, ε is particle emissivity, and σ is the Stefan-Boltzmann constant. C_p was assumed to be the same as graphite, and taken from Merrick [30]. h was obtained by assuming that the Nusselt number was 2 [10]. T_w was 300K. ρ was assumed as 1.4 kg/m^3. ε was assumed as 0.8. σ was 5.67x10^{-8} W/m^{2}K^4.

Figure 12(a) shows time changes of the particle temperature. When particle diameter was small, 0.022 mm, particle temperature increased as soon as gas temperature rose. The gap between the particle temperature and the gas temperature was small. When the particle diameter became large, the rise in temperature became slow. Fig. 12(b) shows time changes of the amount of pyrolyzed volatile matter. Pyrolysis became slow when the particle diameter became large, so that ignition became difficult; i.e. lean flammability limit became large.

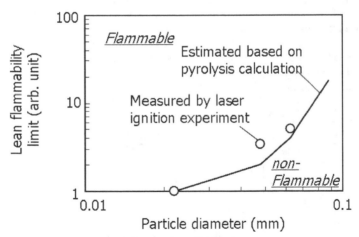

Fig. 13. Effect of particle diameter on lean flammability limit.

Figure 13 shows the relationship between particle diameters and estimated and measured lean flammability limits. Measured lean flammability limit increased with diameter. It was almost proportional to the 1.5th power of the diameter. Estimated results reproduced the tendency.

Recently, development of oxy-fuel combustion technology has been particularly active using pilot-scale plants [14]. Fundamental studies of ignition for oxy-fuel combustion have also been promoted [9, 31, 32]. Suda et al. [9] examined flame propagation velocity under N_2/O_2 and CO_2/O_2 surroundings. When the oxygen concentrations were the same, flame propagation velocities for CO_2/O_2 combustion were lower than those for N_2/O_2 combustion; ignition became difficult for oxy-fuel combustion. Effect of surrounding gas composition should be modeled to allow application of ignition studies to oxy-fuel combustion systems.

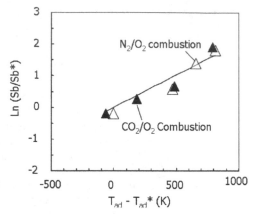

Fig. 14. Relationship between flame propagation velocity and flame temperature formed around the particle [12]. Experimental data were obtained from Suda et al. [9].

The experimental data have been analyzed by paying attention to the flame temperature formed around the particles [12]. Figure 14 shows the relationship between flame temperatures and maximum flame propagation velocities. T_{ad} in the horizontal axis is the adiabatic flame temperatures of volatile matter under the stoichiometry condition [12]. The maximum flame propagation velocity could be predicted from the flame temperature with no dependence on the surrounding gas composition. The relationship between maximum flame propagation velocity and lean flammability limit was already shown in Fig. 8, so that the lean flammability limit could also be predicted from the flame temperature.

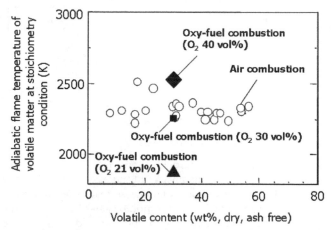

Fig. 15. Calculated adiabatic flame temperatures of volatile matter at stoichiometry condition, under various coals and surrounding gas compositions.

The primary reason why the flame propagation velocity for the CO_2/O_2 combustion was small is that the flame temperature was low, because the specific heat was large. Figure 15 shows adiabatic flame temperature of volatile matter for various coals. When the surrounding gas compositions were the same, effects of coal properties on the adiabatic flame temperatures were small. The adiabatic flame temperature varied when surrounding gas compositions was changed. When oxygen concentrations were the same, the adiabatic flame temperature for oxy-fuel combustion was lower than that of air combustion; ignition for oxy-fuel combustion was difficult. For oxy-fuel combustion, oxygen concentration should be increased to around 30%, in order to secure the same ignition performance as air combustion.

Based on these fundamental results, we developed the model to predict both flame propagation velocity and lean flammability limit [11, 24]. Verification examples are shown in Fig. 16. Calculated results agreed well for various coal properties and experimental conditions.

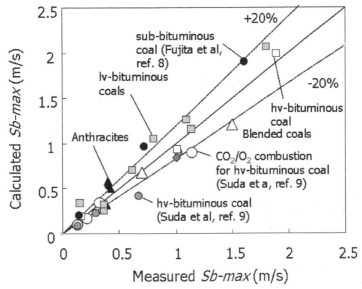

Fig. 16. Verifications of the developed flame propagation model.

3. Lift-off height measurement of pulverized coal jet flame ignited by a preheated gas flow

3.1 Experimental equipment

Lift-off height is usually used for an index of flame stability of continuous gas and liquid flames [33]. This index is used for verification of numerical calculations. However, it has not been examined for coal combustion using simulations. We developed an experimental apparatus to evaluate lift–off height of continuous coal flames, and LES (large eddy simulation) results were verified with the experimental results [23].

Figure 17 shows schematic drawings of the pulverized coal jet flame experiment [27]. The mixture of coal and air was injected through a primary nozzle installed at the centreline of the preheated gas flow and ignited by the surrounding preheated gas. Maximum coal feed rate was 2 kg/h and maximum flow rate of primary air was 3 m³N/h. The nozzle was made of stainless steel and consisted of two concentric tubes with an inner diameter of 7 mm and outer diameter of 25 mm. The injected coals were ignited in the combustion area (open area). Lift-off heights were observed by using a high-speed camera. The surrounding preheated gas was supplied though a honeycomb (100mm square, 30 mm thick) which was symmetrically placed around the nozzle exit. The surrounding gas was formed by catalytic combustion of propane [34]. At first, air was preheated to around 700K by the electrical heater. The preheated air and primary propane mixture was burned in the honeycomb catalyst. The burned gas was mixed with secondary propane downstream form the catalyst. Finally, the gas was heated to 1400-1600K by combustion of secondary propane on the SiC honeycomb. Maximum flow rate of the surrounding preheated gas was 36 m³N/h.

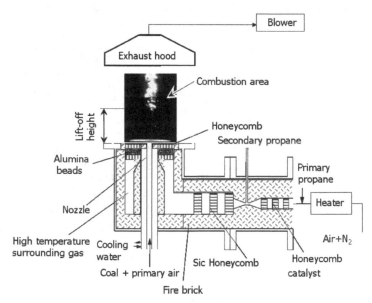

Fig. 17. Schematic drawing of an ignition experiment for lift-off height measurement of continuous flame.

The combustion gas flowed to the outlet through the exhaust hood and the duct. The Reynolds number of the primary jet was about 4500. The inlet stoichiometric ratio (SR) is the most important operating parameter for ignition in this experiment. Downstream from the outlet, a blower vigorously sucked in flue gas. Thus some air was taken from the open area. Prior to the experiment, the amount of the suction flow was estimated from the difference between the amount of the flue gas at the blower and the amounts of the primary air and the preheated gas. The velocity on the side boundary was estimated as 0.16 m/s.

3.2 Flame structure
Figure 18 shows a series of photos showing the ignition process of hv-bituminous coal. Figure 18(a) reproduces instantaneous photos and (b) is the averaged photo. Coal particles were ignited around 150mm downstream from the burner exit [27, 35]. Burning coal

particles were not observed from the burner exit to 150mm downstream. Coal particles were preheated by surrounding gas in this region. Small flames were observed at 150 mm. These small flames grew further downstream, and, formed large flames. A large continuous flame was observed at 200-300 mm. These flame growth phenomena were similar to flame propagation phenomena observed by the laser ignition experiments. The averaged photo showed that the flame luminousness started to increase around 150mm. Lift-off height was defined as the distance between the burner exit and the position that the flame luminousness started to increase.

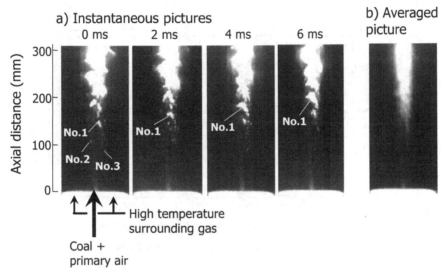

Fig. 18. Example series of flame photos of pulverized coal ignition. Coal particles were ignited by injecting high temperature gas flow.

Figure 19 shows gas temperature, gas composition, and coal burnout profiles along the center axis. Coal concentration in the primary air for Fig. 19 was lower than that for Fig. 18, so that, lift-off height of Fig. 19 was larger than that of Fig.18. Figure 19(a) shows coal burnout profiles. Volatile content of the coal was around 35vol% (dry, ash-free basis). Pyrolysis of fine coals (<0.037 mm) began before ignition. Particle diameter was a very important factor for ignition. The combustible gas formed by pyrolysis of fine particles was strongly related to ignition [23, 27]. Pyrolysis of intermediate size coals (0.037-0.074 mm) began at the ignition region where the small flames formed and grew. Pyrolysis of large coals (>0.074 mm) began after a large continuous flame was formed. The fine particles could strongly contribute to promotion of ignition. The intermediate size particles could also contribute, but, the effect was not large. The large particles could not contribute very much. Figure 19(b) shows gas temperature profiles. A profile when coal was not supplied is shown for comparison. When the luminous flame was formed, the heat of coal combustion was released, so that the gas temperature when coal was present was larger than that without coal. Figure 19(c) shows O_2 and NOx concentration profiles. Oxygen consumption began when the luminous flame was formed. NOx concentration started to increase rapidly after ignition. The rise of NOx is a good index to judge ignition [3].

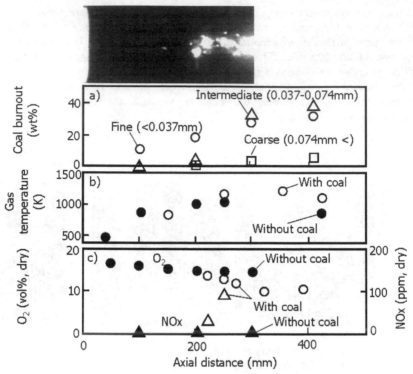

Fig. 19. Axial profiles of coal burnout, gas temperature, O₂ and NOx in the ignition area of coal combustion.

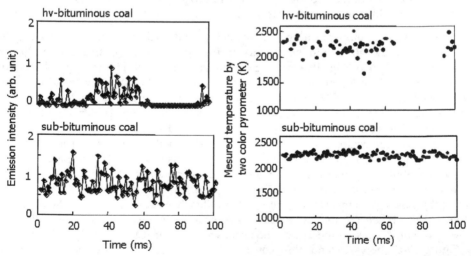

Fig. 20. Particle emission intensity and temperature in the ignition area of hv-bituminous and sub-bituminous coal flames. Particle temperature was measured using a two-color pyrometer.

Particle temperatures (or soot temperature of coal flames formed around the particle) in the ignition region were measured by a two-color pyrometer. The results are shown in Fig.20. Emission intensities and particle temperatures of hv-bituminous and sub-bituminous coals were measured. Emission intensity of sub-bituminous coal was larger than that of hv-bituminous coal. Measured particle temperatures of the two kinds of coals were almost the same. Measured temperatures were around 2200-2300K. These values were almost the same as the adiabatic temperature of the volatile flame formed around the particle (Fig. 15). Averaged gas temperature in the ignition region was around 1000-1300K (Fig. 19). When the coal ignited, particle temperature increased rapidly.

3.3 Relation between the experiments of continuous flames and the laser ignition experiments

In Section 2, we described verification of the flame propagation velocity and lean flammability limit model by laser ignition experiments; these were for unsteady combustion. Then here we verified whether this model could be applied for examination of flame stability for stable flames. Lift-off height is usually used for an index of flame stability. Yamamoto et al. [23] have examined the effect of coal concentration on lift-off height for coal combustion by LES. The flame propagation velocity and lean flammability limit model could successfully evaluate the relationship between coal concentration and flame propagation velocity. They showed that if flame propagation velocity was large, coal could ignite easily.

Lift-off heights were measured for three different primary coal concentrations. Results are shown in Fig.21. The horizontal axis is the calculated flame propagation velocity at the burner exit. A good relationship was observed between calculated flame propagation velocity and lift-off height. When coal concentration increased, flame propagation velocity rose, so that lift-off height became short because coal ignition became easy. Fig. 21 shows results for hv-bituminous coal. A similar conclusion has been obtained for lv-bituminous coal [24]. The flame propagation velocity and lean flammability limit model could also be applied for continuous and stable flames.

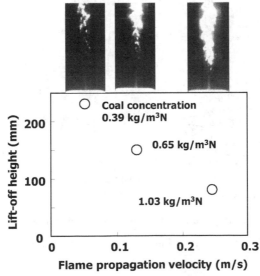

Fig. 21. Effect of primary coal concentration on lift-off height of hv-bituminous coal flames.

4. Application to burner development

In Section 2, we developed the flame propagation velocity and lean flammability limit model by analyzing the result of the laser ignition experiments. In Section 3, the model was applied to examine flame stability of the stable continuous flames. In this section, we paid attention to engineering designs of commercial or pilot scale systems.

Figure 22 is a schematic of the actual boiler and burner arrangement [21]. Blow-off limit fuel concentration of the installed burner is an important design parameter. However, it is not easy to measure the blow-off limits for large scale burners. Development of the model is required to predict the blow-off limits for various burning conditions from limited experimental data.

For actual systems, plural burners are usually installed in one furnace. An example of the burner arrangement is shown on the right side of Fig.22. Heat loss rate from the ignition region of the flames to the furnace wall for actual systems is different from that for small scale equipment. Usually, the heat loss rate for actual systems is small, because one flame is heated by other flames in the neighborhood. Furnace water wall temperature is usually high for actual systems. The influence of radiant heat loss on flame stability should be modeled.

At first, we simulated the influence of radiant heat flux from surroundings to the flame by using the laser ignition equipment. Figure 23 shows the relationship between the radiant heat flux and lean flammability limit. The radiant heat was given by irradiating a continuous laser to the floating pulverized coal particles (Fig.3). The radiant heat flux was regulated by controlling the power of the continuous laser. Usually, the water wall temperature of boilers is 600-700K, and this is equivalent to 1-2 x10⁴ W/m2 of heat flux. The heat flux rose more when one flame received radiant heat from other flames.

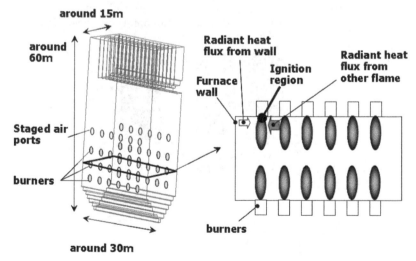

Fig. 22. Example burner arrangement of a pulverized coal fired boiler.

According to Fig. 23, the basic expression of the influence of the radiant heat flux was Eq. (8)[21]

$$1/L = 1/L_0 + a\,Ra \qquad\qquad (8)$$

where L is lean flammability limit, L_0 is lean flammability limit for the standard condition, Ra is radiant heat flux, and a is a constant.

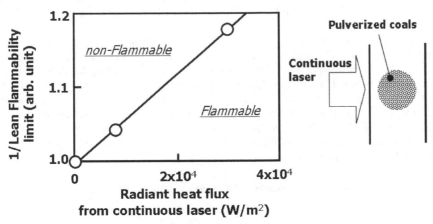

Fig. 23. Effect of radiant heat flux from surroundings to flame on lean flammability limit [21].

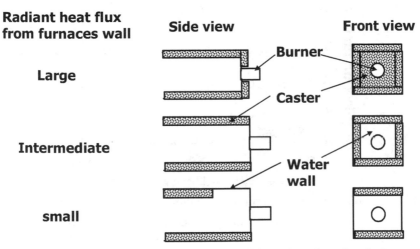

Fig. 24. Example of procedure to evaluate the effect of the radiant heat flux for large scale burners [36].

Figure 24 is an example of the procedure to measure influence of the heat loss by using pilot-scale burners [36]. The pilot-scale burners were almost the same size as for commercial-scale boilers; however, the number of installed burners was smaller. In many case, one large burner was installed for the pilot-scale furnaces. When one burner was installed in the furnace and was surrounded by a water wall, the radiant heat flux from the surroundings was lower than that for commercial-scale boilers. In order to simulate the radiant heat flux for commercial-scale boilers, a part of the furnace wall was surrounded by a caster wall. The effect of the radiant heat flux on flame stability could be evaluated by varying the area of the caster wall. Blow-off limit for commercial-scale boilers could be analyzed from the experimental data by using Eq. (8) [21].

Figure 25 shows the relationship between fuel properties and blow-off limit fuel concentration. The developed model shown in Sections 2 and 3 could predict the effects of fuel properties, particle diameters, and surrounding gas conditions. Blow-off limit (lean flammability limit for actual boiler conditions) could be evaluated by adding the influence of the radiant heat flux. The results of Fig.25 showed that calculated results agreed with experimental results for anthracite, lv-bituminous coals, and hv-bituminous coals [25].

If calculations are verified for several kinds of fuels, blow-off limits for other fuels can be evaluated. Calculated results of blow-off limit for some wood powders are also shown in Fig.25. Biomass fuels are started to be used as an alternative fuel for pulverized coal fired boilers [37]. One of the problems with their use however was that fine grinding was difficult. If the blow-off limit for biomass fuels can be predicted, an appropriate grinding condition can be decided beforehand.

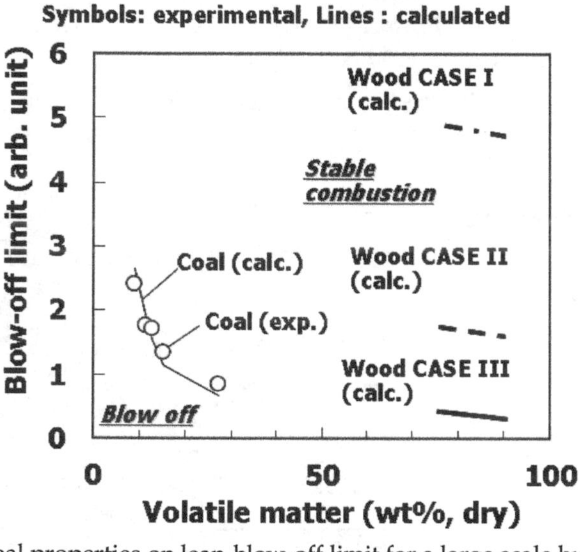

Fig. 25. Effects of coal properties on lean-blow-off limit for a large scale burner.

The effects of diameter on ignition performances for the wood powder were examined in detail. Figure 26(b) shows relationships between fuel concentrations and flame propagation velocities. Figure 26(a) shows diameter distributions for wood powders and hv-bituminous coal used for the case studies. The results for four wood powders and one hv-bituminous coal are shown. Hv-bituminous coal is used widely as fuel for boilers. If flame propagation velocity of the biomass fuel is the same as that of hv-bituminous coal, this biomass fuel is easy to use. Among CASEs I - IV, flame propagation velocity of CASE IV was almost the same as for that of hv-bituminous coal and the diameter distribution of CASE IV was appropriate.

Lift-off height of CASE IV was compared with that of hv-bituminous coal experimentally. The results are shown in Fig. 27. Fuel concentrations were the same. Lift-off height of CASE IV was almost the same level as that of bituminous coal, so that CASE IV was appropriate as the alternative fuel for pulverized coals.

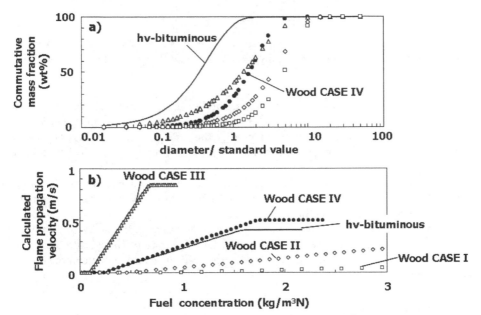

Fig. 26. Relationships between fuel concentration and calculated flame propagation velocities for various coal and wood powders.

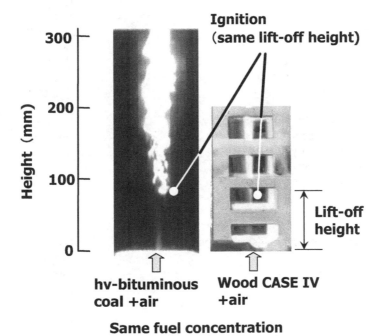

Fig. 27. Comparison of lift-off heights between hv-bituminous coal and wood powder for the same primary fuel concentrations.

5. Conclusion

Application of fundamental ignition experiments for a flame stabilization study was examined for engineering design of coal power plants. A model to evaluate flame stability of small scale experiments and large scale burners was developed based on the fundamental experimental results.

1. A flame propagation velocity and lean flammability limit model for pulverized solid fuels was developed by analyzing fundamental experimental results obtained in laser ignition experiments. The model could predict the effects of fuel properties, particle diameters and surrounding gas compositions on flame propagation velocity and lean flammability limit.
2. The model could be applied for flame stabilization study of coetaneous stable flames. Lift-off height of continuous coal flames could be evaluated from calculated flame propagation velocity by using the model.
3. The model could evaluate blow-off limit of large scale burners. It was found important to evaluate the difference between the effect of heat loss rate of small scale equipment and large scale burners. Case studies were introduced to obtain stable combustion for some biomass fuels.

6. Nomenclature

Av: frequency factor of pyrolysis (1/s)
C_p: specific heat of particles (J/kg K)
D_p : coal particle diameter (m)
Eav_i: average activation energy (kJ/mol)
Ev: activation energy of the pyrolysis (kJ/mol)
$E\sigma_i$: standard deviation of activation energy (kJ/mol)
h: convective heat transfer coefficient (W/m^2 K))
L: lean flammability limit (kg/m^{3}N)
R: gas constant = 8.314 (J/mol K)
Ra: radiant heat flux from surroundings to flame (W/m^2)
S: external surface area of particles (m^2)
Sb: flame propagation velocity (m/s)
$Sb\text{-}max$: maximum flame propagation velocity (m/s)
SR: stoichiometric ratio (-)
T_g: gas temperature (K)
T_p: particle temperature (K)
T_w: wall temperature (K)
V : amount of volatile matter at an instant in the pyrolysis (kg)
V_p: volume of particle (m^3)
V^∞: initial amount of volatile matter (kg)
a_i: constant (-)
b: constant (-)
d: distance between coal particles (m)
n: number of distribution functions of activation energy (-)
s: flame propagation time (s)
t: time (s)

ε: emissivity of particle   (-)
ρ: particle density   (kg/m^3)
σ: Stefan-Boltzmann constant; 5.67×10^{-8} W/m^{2}K^4
Subscripts
i: i th component of volatile matter

7. References

[1] Chen, Y.; et al (1996). *Studying the mechanisms of ignition of coal particles by TG-DTA,* Thermochimica Acta, 275, 149-158.

[2] Tognotti, L; et al (1985). *Measurement of ignition temperature of coal particles using a thermogravimetric technique,* Combust. Sci. Tech., 44, 15-28.

[3] Wall, T. F; et al (1988). *Indicatiors of Ignition for Clouds of Pulverized Coal,* Combust. Flame, 72, 111-118.

[4] Cashdollar, K.L; (1996). *Coal dust explosibility,* Journal of Loss Prevention in the Process Industries, 9, 65-76.

[5] Cashdollar K.L. &. Hertzberg, M; (1983). *Infrared temperature of coal dust explosions,* Combust. Flame, 51, 23-35.

[6] G. E. Andrews and D. Bradley, The burning velocity of methane-air mixtures, Combust. Flame, 19 (1972) 275-288.

[7] Peeters, N; (1999). *The turbulent burning velocity for large-scale and small-scale turbulence,* Journal of Fluid Mechanics, 384, 107-132.

[8] Fujita, O; et al (1993). *HTD-vol. 269, Heat Transfer in Microgravity,* ASME 1993, 59-66.

[9] Suda, T; et al (2007). *Effect of carbon dioxide on flame propagation of pulverized coal clouds in CO_2/O_2 combustion,* Fuel, 86, 2008-2015.

[10] Chen, J. C; et al (1994). *Laser ignition of pulverized coals,* Combust. Flame, 97, 107-117.

[11] Taniguchi, M; et al (1996). *Laser ignition and flame propagation of pulverized coal dust clouds,* Proc. Combust. Inst., 26, 3189-3195.

[12] Taniguchi, M; et al (2011). *Prediction of lean flammability limit and flame propagation velocity for oxy-fuel fired pulverized coal combustion,* Proc. Combust. Inst. 33, 3391-3398.

[13] Ikeguchi, T; et al (2010). *Development of Electricity and Energy Technologies for Low-Carbon Society,* Hitachi Review, 59, 53-61.

[14] Wall, T; et al (2011). *Demonstrations of coal-fired oxy-fuel technology for carbon capture and storage and issues with commercial deployment,* International Journal of Greenhouse Gas Control, (2011) Volume 5, Supplement 1, July 2011, Pages S5-S15, doi:10.1016/j.ijggc.2011.03.014

[15] Ochi, K; et al (2009). *Latest Low-NOx Combustion Technology for Pulverized-coal-fired Boilers,* Hitachi, Review, 58, 187-193.

[16] Iwashige, K; et al (2008). *Numerical simulation for electric power plant systems,* Hitachi Hyoron, 90, 66-71.

[17] Ito, O; et al (2008). *CO_2 reduction technology for thermal power plant system,* Hitachi Hyoron, 90, 20-25.

[18] Handa, M; et al (2008). *Numerical method for three-dimensional analysis of shell-and –tube type of large scale heat exchangers under high temperature circumstances,* Advances in computational heat transfer, CHT-08.

[19] Yamamoto, K; et al (2000). *Development of Computer Program for Combustion Analysis in Pulverized Coal Fired Boilers,* Hitachi Review 49, 76-80.

[20] Yamamoto, K; et al (2005). *Validation of coal combustion model by using experimental data of utility boilers*, JSME International Journal Series B, 48, 571-578.

[21] Taniguchi, M; & Yamamoto, K; (2010). *Fundamental research on oxy-fuel combustion: The NOx and coal ignition reactions*. In: Grace CT, (Ed.), Coal Combustion Research, Nova Science Publishers Inc., New York, pp59-69.

[22] Phillips, S; et al. *Application of high steam temperature countermeasures in high sulfur coal-fired boilers*, available at the address:
http://www.hitachi.powersystems.us/supportingdocs/forbus/hpsa/technical_pa pers/EP2003B.pdf.

[23] Yamamoto, K; et al (2011). *Large eddy simulation of a pulverized coal jet flame ignited by a preheated gas flow*, Proc. Combust. Inst. 33, 1771-1778.

[24] Taniguchi, M; et al (2011). *Application of lean flammability limit study and large eddy simulation to burner development for an oxy-fuel combustion system*, International Journal of Greenhouse Gas Control, Volume 5, Supplement 1, July 2011, Pages S111-S119. doi:10.1016/j.ijggc.2011.05.008

[25] Taniguchi, M; et al (2009). *Comparison of flame propagation properties of petroleum coke and coals of different rank*. Fuel, 88, 1478-1484.

[26] Chen, J. C; et al (1995). *Observation of laser ignition and combustion of pulverized coals*, Fuel, 74, 323-330.

[27] Taniguchi, M. et al (2001). *Pyrolysis and ignition characteristics of pulverized Coal Particles*. ASME Journal of Energy Resources Technology, 123, 32-38.

[28] Niksa S; & C.-W. Lau, C.-W; (1993). *Global rates of devolatilization for various coal types.*, Combust. Flame, 94, 293–307.

[29] Niksa, S; (1995). *Predicting the devolatilization behavior of any coal from its ultimate analysis*, Combust. Flame, 100, 384-394.

[30] Merric, D; (1983). *Mathematical models of the thermal decomposition of coal: 2. Specific heats and heats of reaction*, Fuel, 62, 540-546.

[31] Arias, B; et al (2008). *Effect of Biomass Blending on Coal Ignition and Burnout during oxy-fuel Combustion.*, Fuel, 87, 2753-2759.

[32] Shaddix, C. R; & Molina, A; (2009). *Particle imaging of ignition and devolatilization of pulverized coal during oxy-fuel combustion*, Proc. Combust. Inst., 2091-2098.

[33] Peters, N; (2000). *Turbulent Combustion*, Cambridge University Press, Cambridge, pp. 238-245.

[34] Maruko, S; et al (1994). *Multistage Catalytic Combustion Systems and High temperature Combustion Systems using SiC*, Catalysis Today, 26, 107-117

[35] Taniguchi, M; et al (2000). *Ignition process of pulverized coal Injected into preheated gas flow*, Kagaku Kogaku Ronbunshu, 26, 194-201.

[36] Kiga, T; et al (1987). *Development of IHI Wide Range Pulverized Coal Burner*, Ishikawasjima Harima Giho, 27, 333-338.

[37] Wang, X; et al (2011). *Experimental investigation on biomass co-firing in a 300 MW pulverized coal-fired utility furnace in China*, Proc. Combust. Inst., 33, 2725-2733.

4

Fundamentals and Simulation
of MILD Combustion

Hamdi Mohamed[1], Benticha Hmaeid[1] and Sassi Mohamed[2]
[1]Laboratoire d'Etudes des Systèmes Thermiques et Energétique
[2]Masdar Institute of Science and Technology, Abu Dhabi
[1]Tunisia
[2]United Arab Emirates

1. Introduction

1.1 Introduction and definitions

There is a continual demand to develop combustion systems that lead to reductions in pollutant emissions and increases in energy efficiency, thereby reducing fuel consumption. It is well known that thermal efficiency may be increased by preheating of the reactants (Özdemir & Peters, 2001). An immediate drawback of preheating techniques is that the increased peak flame temperature results in increased NO_x emissions. Alternatively, to reduce NO_x the aim is to reduce peak temperatures, which is often accompanied by an increase in other emissions, for instance carbon monoxide (CO) (Hamdi et al., 2004). It is this conflict in requirements which makes the combination of low emissions, high thermal efficiency burners difficult to design.

A significant development towards combustion which offers both low emissions and high efficiency was found when exhaust gases are recirculated into the reaction zone (Choi & Katsuki, 2001; Hasegawa et al., 1997; Katsuki & Hasegawa, 1998). The heat from the exhaust gases is recovered, increasing system efficiency and also preheating of the reactants. A low oxygen content hot artificial air mixture is obtained where in the exhaust gases act as an inert diluent which basically serves to increase the thermal mass of the system and to reduce fuel consumption. It is this combination of using high temperature and high dilution which is the basis of the Moderate or Intense Low oxygen Dilution (MILD) combustion regime.

It has been suggested that there is an excellent potential for the use of recirculation of exhaust gases to exploit the simultaneous heating and dilution effects to achieve improvements in both the emissions and efficiency, narrowing the gap between the two objectives of low pollutant emissions and fuel savings (Katsuki & Hasegawa, 1998).

In this field, a bewildering number of different terms are used to describe processes which are very similar in nature. While the descriptive titles attempt to highlight some of the subtleties of the various systems, overall many share the same or at least very similar, underlying principles. The sheer number of terminologies which are in use leads to many being misinterpreted or being confused for another. Rather than getting entangled in the twisted web of seemingly endless acronyms and overly descriptive names, for the sake of simplicity, in this discussion the processes will be segregated into one of two groups; preheated air combustion or MILD combustion – which itself being a subgroup of the preheated group.

To some, preheated air combustion refers to a highly specific combustion regime with completely unique and characterisable features. in essence however, the origins and fundamental principle of preheating is the heating of the reactants prior to combustion; which is done by means of recirculation, recuperation or external heating. While there are very important physical differences in the implementation, and in fact in the chemical processes too, preheating will be used to describe any system where the reactants are preheated. Although this may seem an oversimplification, it avoids getting bogged down in the details where this is not necessary for the issue at hand.

Preheating combustion air by means of recovering heat from the exhaust gases is recognised as one of the most effective ways of increasing the thermal efficiency of a combustion system (Borman & Ragland, 1996; Weber et al., 2000; Wünning, J.A. & Wünning, J.J., 1997). The use of such exhaust gas recirculation (EGR) has been developed since the 1970's as a way of increasing furnace efficiency, and subsequently minimising fuel consumption (Weber et al., 2000). The main drawback of increasing thermal efficiency with EGR is there can be an increase in temperatures, leading to increases in NO_x emissions. Depending on the implementation, recirculation of exhaust gases increases the content of inert in the mixture (Wünning, J.A., & Wünning, J.J., 1997).

In the extreme case of significant recirculation of combustion products, where the amount of inert introduced by the recirculation is sufficient to alter the structure of the reaction zone, is where the MILD combustion regime lies. It can therefore be said that there is a close resemblance between preheating processes and MILD combustion. As such, it may be concluded that MILD combustion is a subset of the more general preheating principles, the difference arises from the dilution effects associated with the MILD regime. A further difference between the two is the extent of the preheating. To achieve MILD combustion conditions the temperature prior to combustion must exceed the auto-ignition temperature of the mixture, and is therefore classified as being highly preheated (Choi & Katsuki, 2001; Katsuki & Hasegawa, 1998).

The exact definition of the amount of preheating and recirculation required to move from a simple preheated mode to MILD combustion is somewhat ambiguous, and this is where several different combustion regimes have been identified. One attempt at making a definition of MILD combustion has been made by Cavaliere & De Joannon (2004): "*A combustion process is named MILD when the inlet temperature of the reactant mixture is higher than mixture self-ignition temperature whereas maximum allowable temperature increase with respect to inlet temperature during combustion is lower than mixture self-ignition temperature (in Kelvin)*". Cavaliere and De Joannon (2004) claim that this definition of MILD combustion is unambiguous because the conditions are well defined and univocal – however, it is based on the theory of a well-stirred reactor (WSR) and is therefore not particularly well generalised. An alternative criterion for MILD combustion has been suggested based on the temperature variations in the flow field by Kumar et al. (2002), but no definitive were placed to indicate a general definition.

There are a number of combustion regimes that have been developed which exhibit very similar properties to MILD combustion, each having different names. While each implementation is slightly different, they all operate on very similar principles. Based on the observation that under certain conditions the MILD combustion process may result in no visible or audible flame it has been termed Flameless oxidation (FLOX) by Wünning, J.A. and Wünning, J.J., (1997). On a related theme, some similar names are flameless combustion (Cavaliere & De Joannon, 2004), colourless combustion (Weber et al., 2000) or invisible flames (Choi & Katsuki, 2001). Alternatively, other names have been derived using the requirement of preheating; high temperature air combustion (HiTAC); high temperature combustion technology (HiCOT) (Cavaliere & De Joannon, 2004) or excess enthalpy combustion (Weber et al., 2000). There are subtle differences in each of these combustions technologies, but they all rely on the fundamental principle of the reaction taking place with high dilution levels and in a high temperature environment. For this reason, a single grouping of MILD combustion seems appropriate.

1.2 Principle of MILD combustion
The two basic conditions required for MILD combustion are high dilution and increased temperature of the reactants. Typically, both of these criteria are met by recirculation of the exhaust gases into the reaction zone. Entrainment of exhaust gases into the combustion zone is very important for the initiation of MILD combustion (Özdemir & Peters, 2001). Depending on the characteristics of the recirculation, it is possible to achieve various combinations of recirculation rates and temperatures, resulting in different combustion modes as shown in figure 1 (Wünning, J.A. & Wünning, J.J., 1997).

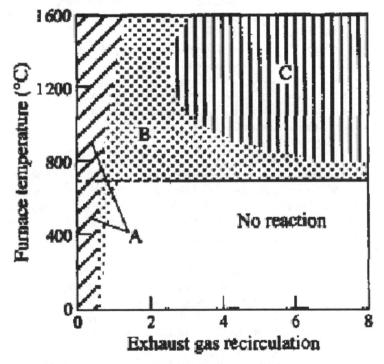

Fig. 1. Stability limits of conventional and MILD combustion (Wünning, J.A. & Wünning, J.J., 1997)

Along the horizontal (x-axis) of Figure1 is exhaust gas recirculation, and is a measure of the proportion of exhaust gas to "clean" air. Mathematically, the recirculation rate may be written as:

$$K_v = \frac{\dot{M}_E}{\dot{M}_F + \dot{M}_A} \tag{1}$$

where $\dot{M}$ refers to mass flow rate, and the subscripts represent; E: recirculated exhaust gas, F: fuel, A: "clean" combustion air. The vertical (y-axis) labelled "Furnace temperature" refers to the temperature of the reactants mixture.

Region A shows the domain of conventional flames, where the recirculation rate is less that~ 30%. Such flames are stable. When the recirculation rate is increased, into region B, the resultant flame has been found to become unstable. If the temperature is below the self-ignition temperature (the horizontal line at ~ 700°C), these flames will extinguish. In region C, it has been found that if the recirculation rate is increased sufficiently, and the temperature is above that of self-ignition, that stable combustion results. It is this mode of operation that is known as the MILD combustion regime.

Figure 1 gives an indication of a fundamental difference between conventional and MILD combustion; the vast increase in the flammability limits (De Joannon et al., 1997), as well as improved flame stability limits (Hasegawa et al., 1997). The reaction at very low oxygen levels, and the improvement in flame stability under MILD conditions, is attributed to the temperature being above that of self-ignition, implying that the flame will always be sustained inside the furnace (Katsuki & Hasegawa, 1998).

By diluting the oxygen stream, the combustion reaction is more distributed, in turn distributing the heat release. With the heat release occurring over a larger volume temperature peaks are avoided, thereby thermal NO formation is largely suppressed (Wünning, J.A. & Wünning, J.J., 1997). Although the heat release occurs over a larger volume, the total is the same, implying that under MILD combustion conditions the heat release rate per unit volume is lower (Hasegawa et al., 1997).

1.3 Characteristics of MILD combustion

One of the most significant effects of dilution under MILD combustion conditions is that the extent of the reaction zone increases. This in turn has the effect of creating a far more uniform temperature distribution throughout the combustion region. In addition to the spatial uniformity of the temperature distribution, there is a significant reduction in the temporal fluctuations too – a reduction in the temporal temperature RMS of 98% has been reported (Hasegawa et al., 1997).

The distributed thermal field associated with MILD combustion also leads to the reduction of peak flame temperatures. By eliminating regions of high temperature the formation of NO_x is largely suppressed. Ten-fold reductions in NO_x emissions are possible with MILD combustion, and no specific region is identified where NO_x is formed (Weber et al., 2000). Although the amount of NO_x produced by MILD combustion is significantly less than conventional combustion systems, some NO_x is still generated. Factors such as composition and mixing can affect the local combustion properties and thus the amount of NO_x produced. For instance, the location of the inlet nozzle (hence the mixing process) drastically affects NO_x emissions, and cannot be attributed to imperfect combustion since no unburned hydrocarbons are measured in the exhaust (Choi & Katsuki, 2001). Different diluents also influence the production of NO_x, with much less NO_x being formed with CO_2 dilution as compared to N_2.

Despite the less intense reaction surrounding MILD combustion, both unburned hydrocarbons and carbon monoxide (CO) level are very low, or even not detectable, at the outlet of a MILD combustion furnace (Weber et al., 1999, 2000). The low emission of CO (< 100 ppm) suggests stable combustion conditions exist with such furnaces (Hasegawa et al., 1997). The temperature range encountered in MILD combustion also has significant advantages in the reduction of soot (Cavaliere & De Joannon, 2004).

To achieve stable combustion at the low oxygen levels associated with the very high recirculation rates of MILD combustion it is necessary for the initial temperature to exceed the auto-ignition temperature (Katsuki & Hasegawa, 1998). The method of achieving recirculation can be internal or external with regard to the combustor (Szegö et al., 2007). The recirculation, typically achieved with high velocity nozzles (Cavaliere & De Joannon, 2004), and the resultant reduction in oxygen concentration, lead to MILD combustion flames being associated with low Damköhler numbers (Katsuki & Hasegawa, 1998). In conventional flames, the chemical kinetics occurs much faster than the mixing, whereas under MILD conditions, the combustion is controlled by both the kinetics and the mixing (Milani & Saponaro, 2001). As an extension to this, in a furnace environment, MILD combustion may be likened to a well-stirred reactor (WSR) as the chemical time scales becoming larger in relation to the turbulence time scales (Plessing et al., 1998). The link between MILD combustion and a well Stirred Reactor may tend to indicate that this regime should be readily modelled, since a well stirred reactor is well defined. The interaction between the turbulence and the chemistry in describing the nature of MILD combustion is expected to play a significant role however (Katsuki & Hasegawa, 1998), and so the analogy to a WSR may not always be applicable.

One of the most noticeable characteristics of MILD combustion is the differences in visual appearance compared to other flames. When, operating in MILD conditions, furnaces have been described as "glowing", with no actual flame visible (Weber et al., 2000). This type of description of the appearance of MILD combustion is very typical. MILD combustion is frequently described as appearing weak, with diffuse and distributed reactions zones (Katsuki & Hasegawa, 1998).

The visible emission from MILD combustion is measured to be two orders of magnitude lower than from conventional flames (De Joannon et al., 2000). In the case of methane, the predominate sources of colour in typical flames is due to C2 and CH, both of which are formed at relatively high temperatures, which are avoided in MILD conditions, therefore the lack of colour (Cavaliere & De Joannon, 2004). Descriptions of combustors typically describe a very weak, bluish coloured flame which is barley visible (Kumar et al., 2002). The colour of the flame is also dependent on the type of fuel, with LPG fuel possibly resulting in a blue/green colour (Hasegawa et al., 1997), but this is dependent on the diluent (Cavaliere & De Joannon, 2004).

2. Application and implementation of MILD combustion

Driven by the industrial relevance, the MILD combustion process has so far predominately been implemented and investigated in furnaces. In most combustion systems, combustion efficiency is sufficiently high, but there is room for improvement in terms of heat utilization (Choi & Katsuki, 2001). Economical benefits are sought by increasing thermal efficiency, which can be accomplished by using preheating, and can achieve fuel savings up to 60% (Weber et al., 1999).

Large amounts of waste heat can be recovered in furnaces using heat-recovery devices to preheat the combustion air, typically above 1300 K (Katsuki & Hasegawa, 1998). To achieve preheating, energy from the exhaust gases is transferred back to the combustion air either by recuperative or regenerative heat exchangers (Plessing et al., 1998).

As an extension to using preheated air to achieve energy savings, dilution with exhaust gases has been seen to lower NO_x emissions, although the low O_2 levels require even higher preheat temperatures than generally used for previous heat recovery systems (Choi & Katsuki, 2001). Using highly preheated air (> 1100 K), in conjunction with low O_2 levels (<15%), can accomplish energy savings, improved thermal performance and lower NO_x emissions (Hasegawa et al., 1997). Worth highlighting is that although the oxygen levels used in MILD combustion are very low, on a global level, furnaces are operated lean (Kumar et al., 2002).

Preheating can also be achieved internal to the combustor by means of internal flow recirculation to provide heating of the unburnt gases (Kumar et al., 2002; Plessing et al., 1998). Due to the high temperatures, self-ignition in MILD combustion conditions is assured and so high-velocity jets can be used (Cavaliere & De Joannon, 2004). The use of high velocity jets is advantageous to ensure adequate mixing of the fuel, air and recirculated exhaust gases is achieved to simultaneously heat and dilute the reactants (Plessing et al., 1998). Such internal recirculation is the key principle of the FLOX burners which are at the forefront of the industrial application of MILD combustion (Wünning, J.A. & Wünning, J.J., 1997).

The key features of furnaces operating in the MILD combustion regime are the increased thermal efficiency, flat thermal field and lower NO_x emissions. In relation to a conventional combustion chamber, the radiation flux is higher in the first zone on the combustor and lower in the second part, which overall leads to a more uniform thermal field (Cavaliere & De Joannon, 2004). In the work of Kumar et al. (2002), they found that the normalised spatial temperature variation is 15% for MILD, compared to 50% for their classical jet flame. Moreover, the heat flux uniformity achieved through MILD combustion is a highly desirable feature that often cannot be met with conventional burner technology (Hasegawa et al., 1997). The heat flux of a furnace operating in MILD regime is almost constant heat flux, which may provide high heat transfer rates, and is very desirable for a number of industrial processes (Weber et al., 2000). It is estimated that MILD combustion technologies are capable of a 30% energy saving (Katsuki & Hasegawa, 1998). Although achieving greater efficiency can be achieved using MILD combustion, improved product quality as a result of the thermal field is also a major driving force behind the process (Cavaliere & De Joannon, 2004).

The lack of flame and lower peak temperatures in MILD combustion may be thought to be less efficient at heat deposition, but because of the larger reaction volume the total heat radiation flux can actually be higher in MILD combustion than in conventional flames (Weber et al., 2000). This has been shown previously; where radiation emitted downstream from the reaction zone is about five times that emitted upstream (Williams, 1985), implying that although the peak temperature within MILD combustion is lower there is significant radiation emitted. Moreover, while chemiluminescence is low in MILD combustion, and may provide flames with their distinctive colours, the associated energy transfer by this mechanism is negligible in comparison with radiation from major stable compounds (such as H_2O and CO_2) (Williams, 1985)..As an example of the increased efficiency offered by MILD combustion, Wünning, J.A. and Wünning, J.J. (1997) used CFD modelling to determine the input energy required to achieve 160 kW output from a furnace. Without any preheating, the burner capacity was 400 kW, with 600 °C preheat using a recuperator the input was 245 kW (Wünning, J.A. & Wünning, J.J., 1997). For FLOX mode, with 950°C preheat the required input was as low as 200 kW – half that of the no preheat case (Wünning, J.A. & Wüninng, J.J., 1997). Furthermore, the peak production rate of NO_x is reduced by several orders of magnitude (Wünning, J.A. & Wünning, J.J., 1997). This highlights why industry is seeking this technology to deal with reducing fuel costs and meeting more stringent NO_x emission targets (Brggraaf et al., 2005).

A schematic layout of a MILD combustion furnace is shown in Figure2 from Choi & Katsuki (2001). This furnace is based on an alternating flow system, and comprises of two sets of regenerators. In one flow direction, the incoming air passes over a hot ceramic honeycomb, which preheats the incoming air. As the hot exhaust gases leave the furnace at the other end they pass through another ceramic honeycomb, and subsequently heat it. After a certain period of time, the temperature of the incoming honeycomb decreases, whilst that of the exhaust side honeycomb has increased, and the flow direction is then reversed. The now hot honeycomb acts to heat the incoming air, whilst the cooler honeycomb is now heated by the exhaust gases. The process is then again reversed and continues the alternating heating and cooling of the regenerators.

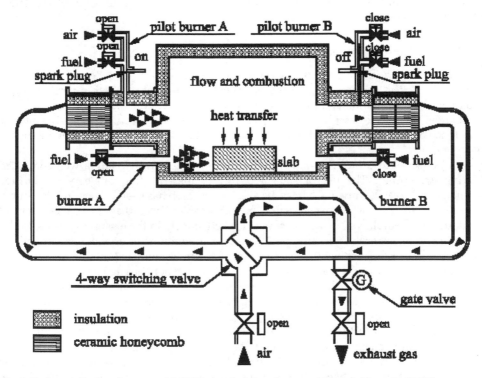

Fig. 2. Industrially implemented MILD combustion furnace (Choi & Katsuki, 2001)

In addition to furnace design, MILD combustion has potential for inclusion in other practical application too. The implementation used for furnaces clearly require significant alteration for use in other applications, but, at least in principle, the underlying aspects of

MILD combustion can be extended to other fields. The low temperature and uniform thermal fields are advantageous attributes in the design of gas turbines, particularly from the materials standpoint (Cavaliere & De Joannon, 2004). Advancement in the application of some aspects of MILD combustion to the design of gas turbines has been made, but issues with self-ignition and dilution has meant generalised application has not yet been achieved (Cavaliere & De Joannon, 2004).

Along a similar vein, some developments which are similar to MILD combustion have been applied to diesel engines but, to date these are not what could be considered a fully fledged MILD combustion system (Angrill et al., 2000; Choi & Katsuki, 2001). An engine which is much closer to practical implementation of MILD combustion is the Homogenous Charge Compression Ignition (HCCI) engine (Cavaliere & De Joannon, 2004).

The preheating of combustible mixtures, which is central to MILD combustion, offers the ability to use fuels that would not typically be used - in particular, low calorific value fuels (Choi & Katsuki, 2001; Katsuki & Hasegawa, 1998).

There is clearly potential for MILD combustion to offer significant benefits in a number of applications. One of the biggest impediments related to the implementation of this promising technology is the lack of fundamental knowledge of the reaction zone structure. A better understanding of the flame structure is needed to see widespread implementation of MILD combustion (Muruta et al., 2000).

3. MILD combustion studies

As was highlighted in the preceding section, MILD combustion has potential for use in a number of practical applications, but this is limited by a lack of detailed understanding of fundamental aspects. Although previous research has been conducted in the area of MILD combustion, this work has predominately concentred on large scale systems. Despite the importance of these systems, such methodology fails to address the fundamental issues of the regime.

This project seeks to differ from much of the existing research by using a well controlled experimental burner. The use of such a burner enables a wide range of combustion parameters to be easily varied, whereas the existing research has been limited in this respect. In addition, the use of an experimental burner enables laser diagnostic measurement techniques to be employed, which cannot be used effectively in the existing systems. In this way, utilizing an experimental burner will avoid the limitations of previous investigations, whilst simultaneously enabling the required conditions to be emulated.

Although much of the previous study of MILD combustion has tended to be directed towards industrial systems, there have been some studies on the fundamental aspects. Due to the lack of visible flame, it has been necessary to resort to non-luminosity based techniques. As an example, Plessing et al. (1998) used laser induced predissociative fluorescence (LIPF) of OH to visualise the reaction zone. In conjunction with the OH-LIPF, Rayleigh measurements were also recorded. Rayleigh scattering is particularly suitable for MILD combustion systems as the high levels of dilution result in the Rayleigh cross-section varying by less than 2% between the burnt and unburnt gases (Plessing et al., 1998), further enhanced by the lack of particulate matter.

The furnace from Plessing et al. (1998), has been used in further studies, including Özdemir & Peters (2001), Coelho & Peters (2001) and Dally et al. (2004). Özsemir & Peters (2001) extended the OH and temperature measurements of the furnace by including flow field measurements, finding that strong shear is an important criteria for attaining the high mixing rates necessary to create MILD combustion conditions. Coelho & Peters (2001) numerically simulated the same furnace and compared the results to the previous measurements. The models qualitatively matched the experimental results, although differences were noted near the burner exit and also in NO formation. The same furnace has also been used by Dally et al. (2004) to study the effects of fuel dilution, both numerically and experimentally. They found that the shift in stoichiometry caused by fuel dilution helped in the establishment of MILD combustion.

Other furnace designs have been studied both experimentally and numerically by Weber et al. (1999, 2000) and subsequently modelled by Mancini et al., (2002). Similar to Coelho & Peters (2001), Mancini et al. (2002) found that simulations matched the experiment finding, except near the jet.

On a more fundamental level, Dally et al. (2002a, 2002b) reported on the structure of hydrocarbon non-premixed laminar and turbulent flames stabilised on a jet in a hot and diluted co-flow (JHC) burner. They used single-point Raman-Rayleigh-LIF diagnostic techniques to simultaneously measure temperature, major and minor species at different locations in these flames. They found that major changes in the flame structure occur when reducing the oxygen concentration and that, at higher jet Reynolds number and low oxygen concentration, oxygen leakage from the surrounding may cause local extinction of the flame front.

As an extension to the experimental work of Dally et al. (2002a, 2002b), CFD modelling has subsequently been applied to the same flame conditions (Christo & Dally, 2005; Kim et al., 2005). In these CFD studies, the models have shown that the experimental results can be reproduced, albeit with significant deviations in some situations, furthermore, only some of the model variants gave good agreement. In all cases, obtaining reliable CFD models proved most difficult at the very low O_2 conditions, and at the downstream measurement locations where the entrainment of surrounding air introduces an additional degree of complexity. Overall, these modelling studies indicate that there is potential for application of CFD to MILD combustion conditions, but some further work is still required.

In a study of a laboratory MILD combustion burner it has been shown that reactions under the MILD combustion regime occur in disconnected zones where the OH and temperature intensities differ (Dally & Peters, 2002; Plessing et al., 1998).

The explanation for this discontinuity between the reaction and the temperature is attributed to the varying amounts of recirculated exhaust gases, which change the balance required for MILD combustion. The conclusion of the work of Plessing et al. (1998) is that MILD combustion resembles a reaction in a well-stirred reactor, although further investigations are necessary for complete analysis of MILD combustion.

It has been suggested that a furnace operating under MILD conditions is similar to a well-stirred reactor (WSR) (Weber et al., 1999). Based on the concept of attempting to infer MILD combustion from a well-stirred reactor, De Joannon et al. (2000), has attempted to model a WSR with MILD combustion conditions. This work did not compare the composition of a well-stirred reactor to any experimental data, but did examine the effect of the diluent composition. They found that there is a dramatic effect on the combustion composition, depending on the composition of the diluent. Also the effect of the residence time on the reaction was of no significance for the majority of the temperature range 500 – 2000 K. However as identified in their work, De Joannon et al. (2000), acknowledge that a well-stirred reactor is unfeasible for a practical combustor.

4. The potentialities of the MILD combustion in application related to the pollutant emission

Several works exploit the influence of MILD combustion on formation of NO_x through the different reaction paths. It is well known that nitrogen oxides form along there possible paths, namely thermal NO_x, fuel-NO_x and prompt NO_x. The first two mechanisms, which are more efficient in the NO_x formation, are depressed by the low-temperature in MILD combustion. The first mechanism is most effective for high levels of temperature and oxygen, according to the Zel'dovich mechanism, whereas the second one is maximized for high levels of fuel richness and fuel nitrogen content. The last mechanism relies on high reactant temperature also and can proceed in very rich conditions only if the temperature is high enough to sustain the process.

Wünning, J.A. & Wünning, J.J. (1997) reviewed flameless oxidation using experimental and numerical techniques. The study reveals the ability of flameless combustion to lower NO_x emissions through lowered adiabatic flame temperatures. The study, however, used low inlet air temperatures and did not include a reaction flow analysis study.

Ju & Niioka (1997) studied NO formation in a non-premixed two-dimensional laminar jet flame, another canonical configuration for the understanding of the effects of turbulence on combustion. The study shows that the formation of prompt NO is observed in the flame reaction zone, while the thermal NO mechanism is the predominant path for NO formation on the high-temperature air side. The emission index of NO was shown to have a high sensitivity to the oxidizer preheat temperature, increasing drastically with an increasing preheat temperature.

Hamdi et al. (2004) investigated the high temperature air combustion of partially premixed (lean or rich) flames (methane/air) in counter-flow geometry (figure 3). The study illustrated that the flame structure for the rich partially premixed flame is significantly affected by the oxygen concentration in the air and a significant reduction in NO_x emission is observed. However,

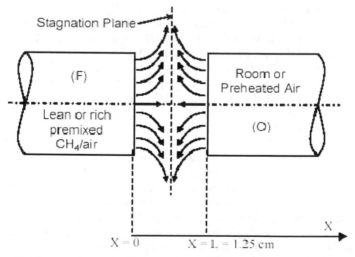

Fig. 3. Opposed jet flame schematic; Preheat air vs. Lean or rich premixed methane/air (Hamdi et al., 2004)

Figure 4 reports several accumulated NO$_x$ data in a log scale as a function of process temperature assuming very efficient preheating of the combustion air (60-80% of the process temperature).

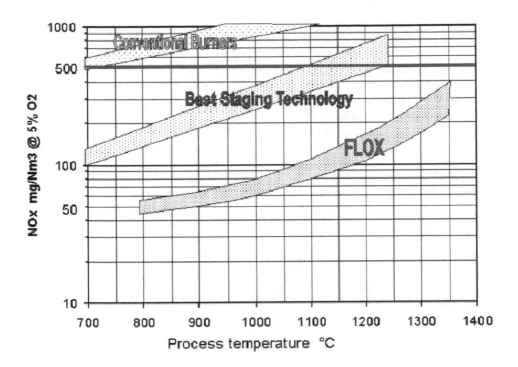

Fig. 4. NO$_x$ emissions vs. process temperature (Milani & Saponaro, 2001)

It may be seen from Figure4 that MILD combustion (region labelled FLOX) may abate NO$_x$ emissions by one order of magnitude even with respect to the best staging techniques for Low-NO$_x$, envisaged for natural gas firing. The main reason for this excellent result stem from the well known circumstance, that thermal NO formation is extremely sensitive to flame temperature peaks or spikes and these are now cut away in MILD combustion firing. But also the other known NO formation mechanisms are positively modified, as prompt NO depends on radicals that are abundant in a flame front, but much reduced in MILD combustion mode (Plessing et al., 1998) and also fuel NO may undergo reburning effects capable of reconverting NO into N$_2$ species (Eddings & Sarofin, 2000; Weber et al., 2000).

Recently, Hamdi et al. (2009) studied the effects of fuel dilution and strain rate on zone structure and NO$_x$ formation in MILD combustion. The study is based on parametric simulations of non-premixed counter-flow flames of methane with nitrogen dilution and highly preheated air.

A counter-flow laminar flame, therefore, is the best field if we want to discuss the local combustion characteristics of the industrial furnace, in particular, using detailed chemical kinetics. Although the real counter-flow flame is inherently two-dimensional, the assumption and formulation of a similarity solution (Giovangigli & Smooke, 1987; Kee et al., 1987; Darabiha et al., 1988) are applied to transform the problem into one-dimensional. Therefore, one-dimensional solutions along x axis represent the whole field of the flat flame formed perpendicular to x axis by the similarity. The governing equations of the flow can be found in refs. (Giovangigli & Smooke, 1987; Kee et al., 1987; Darabiha et al., 1988).

The solution is computed using Sandia's OPPDIF code (Lutz et al., 1996) along with the GRI-Mech 3.0 mechanism (Glassman, 1996) for methane and NO_x chemistries.

Through their study, it can be seen in figure 5, that there is a general decrease in the production of NO and H_2 with a decrease in fuel concentration. This is due to the reduced fuel present to transport to the flame.

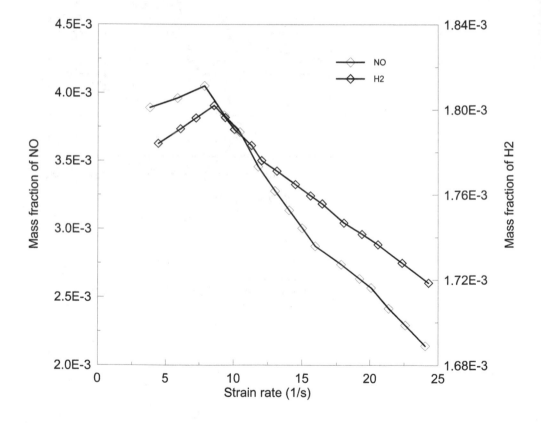

Fig. 5. Mass fractions for NO and H_2 vs. strain rate at air preheated temperature of 2000K and fuel concentration of 15%

The emission index of NO was found to decrease as fuel concentration is lowered due in part to the weakening reaction as fuel concentration is decreased, as shown in figure 6.

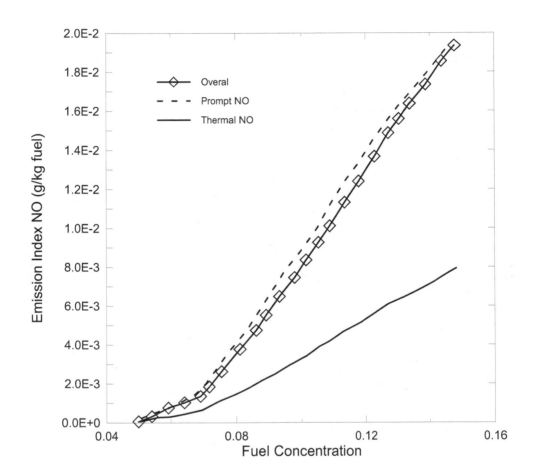

Fig. 6. Emission index profiles for NO_x vs. fuel concentration at air preheated temperature of 1000K and strain rate of 4s[-1]

While there is a decrease in the overall contribution of both thermal and prompt NO as fuel concentration is decreased, there is also a decrease in the individual contributions of thermal and prompt NO, as indicated in figure 7. Due to the high dependence on temperature seen in the thermal NO mechanism, the thermal NO mechanism is dominant at higher preheat temperatures, while the prompt is dominant at lower preheat temperatures.

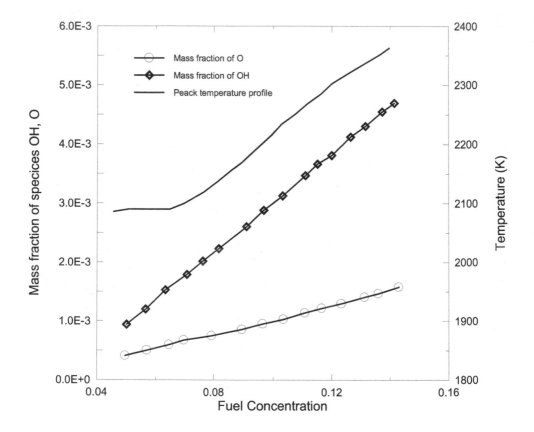

Fig. 7. Mass fractions of O and OH vs. fuel concentration at air preheated temperature of 2000 K and strain rate of 24 s^{-1}

Hamdi et al. (2009) found, in contrast with fuel concentration, that strain rate has been seen to have a large effect on the slower mechanisms, but not the fast ones due to a balance between decreased residence times and higher transport and vice versa, as illustrated by figure 8.

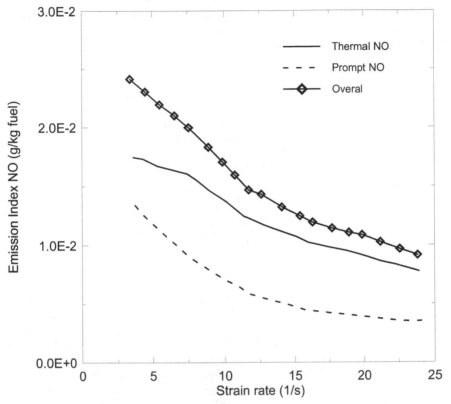

Fig. 8. Emission index profiles for NO_x vs. strain rate at air preheated temperature of 2000 K and fuel concentration 5%

5. Turbulence-chemistry interaction in MILD combustion

Modeling the interaction of a complex set of chemical reactions and high Reynolds number turbulent fluid flow typical for gas turbine combustion is a challenging task. If the emphasis is on the formation of pollutants such as NO_x a full chemistry model is prohibitive. Direct numerical simulation is not feasible due to high computational costs and storage requirements. Even a probability density function (PDF) modeling approach of a "real" gas turbine combustion chamber with boundary effects, swirling inhomogeneous flow by far exceeds computational capacities. Therefore, severe simplifications have to be made to be able to study the interaction of turbulence and detailed chemistry with respect of NO_x formation. In the MILD combustion model, the combustor was simulated by the flow model shown in figure 9; the hot gases exit the first combustor chamber and enter the second combustor chamber.

The first combustion chamber is assumed to be an ideal, turbulent, adiabatic, constant pressure, well stirred reactor (PSR). It is assumed there are no boundary effects and the turbulence created at the inlet is homogeneous, isotropic, and stationary. Then, the second combustion chamber is described as a "Partially Stirred Reactor (PaSR)", where mixing and chemical reactions occur simultaneously. The widely available CHEMKIN package (Miller et al., 1996) and specially its software application PaSR is used to model the flow field that occur in the 2nd combustion chamber.

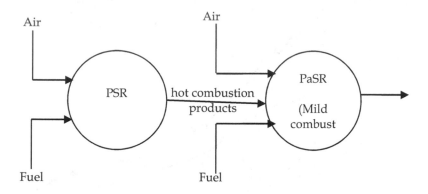

Fig. 9. Principle of the MILD combustion model

5.1 Reactor equations

The Partially Stirred Reactor (PaSR) is an extension of a Continuously Stirred Tank Reactor (CSTR) model that addresses the interaction between chemical reactions and turbulence (Chen, 1997; Correa, 1993). The basic assumptions for the PaSR are similar to those of the CSTR or PSR (perfectly Stirred Reactor). The major difference between a PSR and a PaSR lies in the treatment of the molecular mixing inside the reactor. In a PSR the contents of the reactor are well mixed by assuming high-intensity turbulent stirring action and the only influence from fluid dynamics is controlled by the reactor residence time τ_R. Unlike the PSR, a PaSR allows fluid dynamics to control the extent of the mixing and consequently the chemical reactions by means of an additional parameter: the scalar mixing time, τ_{mix}. The turbulent mixing time scale is often considered to be proportional to the turbulent eddy turnover time as (Kee et al., 2002):

$$\tau_{mix} = C_D \frac{\tilde{k}}{\tilde{\varepsilon}} \qquad (2)$$

Where C_D usually treated as a constant but its value varies for different flow configurations, as suggested by Pope (1991). C_D is set $C_D = 2.0$. The ratio of turbulent kinetic energy to its dissipation rate, $\tilde{k}/\tilde{\varepsilon}$, represents the time scale of the energy-containing eddies which characterize the turbulent mixing action.

The composition and temperature in the PaSR are described by a probability density function (PDF). This composition PDF is a subset of the joint velocity-composition PDF because the flow field in the PaSR is assumed to be spatially homogeneous. Velocity fluctuations are also ignored; that is, the PDF is over scalars only, but is not a delta-function in scalar space because reactants, intermediates, and products are not mixed at the molecular level.

The PaSR consists of an adiabatic chamber having M inlet streams and one outlet. Steady flows of reactants are introduced through the inlets with given gas compositions and temperatures. The reactor pressure is assumed to be constant.

The overall mass balance for the gas mixture inside the PaSR is:

$$\frac{d(\prec \rho \succ V)}{dt} = \sum_{i=1}^{M} \overset{\bullet}{m}_i - \overset{\bullet}{m}_0 = 0 \tag{3}$$

where $\overset{\bullet}{m}_i$ is the mass flow rate of the ith inlet and $\overset{\bullet}{m}_0$ is the through-flow mass flow rate.

The average properties of the PaSR are obtained from the ensemble of particles inside the reactor. Each particle is regarded as an independent PSR and interacts with others only trough the molecular mixing process. Therefore, the conservation of energy and species is applied to an individual particle rather than to the reactor.

The species equation for a particle is then similar to that of a PSR:

$$\frac{dY_k^{(n)}}{dt} = \frac{1}{\overset{\bullet}{m}_0 \tau_R} \sum_{i=1}^{M} \left\{ \overset{\bullet}{m}_i \left(Y_{i,k} - \prec Y_k \succ \right) \right\} + \frac{W_k \overset{\bullet}{\omega}_k^{(n)}}{\rho^{(n)}} \tag{4}$$

and so is the energy equation for a particle is:

$$\frac{dT^{(n)}}{dt} = \frac{1}{C_p^{(n)} \overset{\bullet}{m}_0 \tau_R} \sum_{i=1}^{M} \overset{\bullet}{m}_i \left(\sum_{k=1}^{k_g} Y_{i,k} \left(h_{i,k} - \prec h_k \succ \right) \right) - \sum_{k=1}^{k_g} \frac{W_k \overset{\bullet}{\omega}_k^{(n)} h_k^{(n)}}{\rho^{(n)} C_p^{(n)}} \tag{5}$$

In the above equations, the angled bracket (< >) indicates the ensemble average that we use to approximate the density-weighted average in the simulation. The average residence time of the reactor, τR, is calculated as:

$$\tau_R = \frac{\prec \rho \succ V}{\overset{\bullet}{m}_0} \tag{6}$$

In the following section we discuss the influence of the turbulent mixing on the chemical kinetics under MILD combustion conditions. In order to demonstrate the influence of turbulent mixing on the chemical reactions we vary the turbulent mixing time τmix. From the limiting case of the PSR (i.e. τmix = 0s), which corresponds to very fast mixing, we move to realistic turbulent time scales τmix = 50 ms, which is slow mixing and an intermediate mixing time τmix = 25 ms.

The computational results show, through figure 10, that the combustion processes as well as the NOx formations are very sensitive on the mixing intensity. With increasing turbulent mixing time (i.e. decreasing mixing intensity) the combustion process is stretched out. The ignition delay is shorter but the residence time to achieve complete combustion increase significantly.

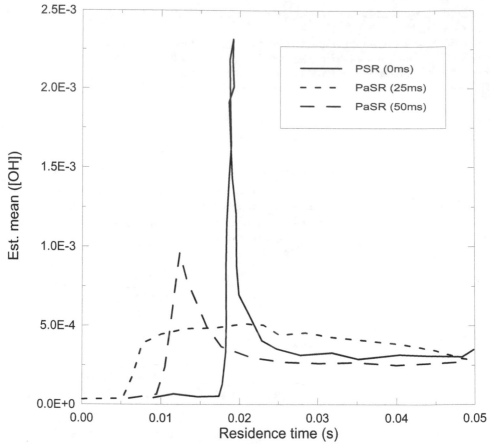

Fig. 10. Time evolution of estimated mean of OH mass fraction in the PSR and PaSR cases

6. Conclusions

This review chapter collects information which could be useful in understanding the fundamentals and applications of MILD combustion. The pieces of information in this field are still sparse, because of the recent identification of the process, so that many speculative considerations have been presented in order to make the whole framework more consistent and rich with potential new applications.

The main points to be stressed as concluding remarks pertain to three main considerations. The first is that MILD Combustion has to be considered a new combustion regime. It is a combustion process that achieved a desirable combination of low pollutant emissions and improved fuel savings.

The second consideration is that MILD combustion can fin a possible application in fuel reforming to produce hydrogen. However the three parameters (fuel concentration, strain rate and oxidizer preheat temperature) discussed in the study described in Section 4, play major roles in the balance between the desired product (hydrogen) and the undesired by-product (NO).

The third consideration is that the characteristic times of kinetics and turbulence, for the MILD combustion regime, become comparable and the two phenomena are coupled with each other. It is therefore of great interest to determine the interaction of the turbulent mixing and the combustion process as well as its impact on NO_x emission. This study has been described in Section 5. A Partially Stirred Reactor model has been developed to study the influence of turbulent mixing intensity on MILD combustion and NO_x formation in gas turbine combustor.

Nomenclature

C_D	Model constant (= 2.0)
C_P	Specific heat at constant pressure [J/kg K]
h_k	Specific enthalpy of species k [J/kg]
K_V	Recirculation rate [-]
$\tilde{k}$	Turbulent kinetic energy [m²/s²]
k_g	Total of species included in the gas-phase
$\dot{M}$	Mass flow rate [kg/s]
$\dot{m}_0$	Through-flow mass flow rate [kg/s]
$\dot{m}_i$	Mass flow rate of the ith inlet [kg/s]
T	Temperature [K]
t	Time [s]
V	Reactor volume [m³]
W_k	Molecular weight of species k [kg/mol]
Y_k	Mass fraction of species k [-]

Greek scripts

$\tilde{\varepsilon}$	Dissipation ratio of turbulent energy [m²/s³]
ρ	Density [kg/m³]
τ_{mix}	Turbulent mixing time [s]
τ_R	Reactor residence time [s]
$\dot{\omega}_k$	Reaction rate of species k [s⁻¹]

Subscripts

A	Clean combustion air
E	Recirculated exhaust gas
F	Fuel
i	i^{th} inlet
k	k^{th} species
O	Oxidizer

7. References

Angrill, O., Geitlinger, H., Streibel, T., Suntz, R. & Bockhorn, H (2000), Influence of Exhaust Gas Recirculation on Soot Formation in Diffusion Flames, *Proceedings of the Combustion Institute*, Vol. 28, pp. 2643-2649.

Borman, G.L. & Ragland, K.W. (1996), *Combustion Engineering*, International Ed, The McGraw-Hill Companies, Inc.

Burggraaf, B.T., Lewis, B., Hoppesteyn, P.D., Fricker, N., Santos, S. & Slim, B.K. (2005), Towards Industrial Application of High Efficiency Combustion, *International Flame Research Foundation*, Research Digest IFRF Doc. N° K70/y/156.

Cavaliere, A. & De Joannon, M. (2004), Mild Combustion. *Progress in Energy and Science*, Vol. 30, pp. 329-366.

Chen J. Y. (1997), Stochastic Modeling of Partially Stirred Reactors. *Combustion Science and Technology*, Vol. 122, pp. 63-94.

Christo, F.C. & Dally, B.B. (2005), Modeling Turbulent Reacting Jets Issuing into a Hot and Diluted Co-flow. *Combustion and Flame*, Vol. 142, pp. 117-129.

Choi; G. M. & Katsuki, M. (2001), Advanced low-NO_x Combustion using Highly Preheated Air. *Energy Conversion and Management*, Vol.42, pp. 639-652.

Coelho, P.J. & Peters, N. (2001), Numerical Simulation of a Mild Combustion Burner. *Combustion and Flame*, Vol. 124, pp. 503-518.

Correa S. M. (1993), Turbulence-Chemistry Interactions in the Intermediate Regime of Premixed Combustion. *Combustion and Flame*, Vol. 93, pp. 41-60.

Dally, B.B., Riesmeier, E. & Peters, N. (2004), Effect of Fuel Mixture on Moderate and Intense Low Oxygen Dilution Combustion. *Combustion and Flame*, Vol. 137, pp. 418-431.

Dally, B.B., Karpetis, A.N. & Barlow, R.S. (2002a), Structure of Jet Laminar Non-premixed Flames under Diluted Hot Co-flow Conditions, *Australian Symposium on Combustion and the 7th Australian Flame Days*, Adelaide, Australia. (2002)

Dally, B.B., Karpetis, A.N. & Barlow, R.S. (2002b), Structure in Turbulent non-premixed Jet Flames in a Diluted Hot Co-flow, *Proceedings of the Combustion Institute*, Vol. 29, pp. 1147-1154.

Dally, B.B. & Peters, N. (2002), Effect of Fuel Mixture on Mild Combustion, *Work done during sabbatical* at ITM in 2002.

Darabiha, N., Candel, S.M., Giovangigli, V. & Smooke, M.D. (1988), Extinction of Strained Premixed Propane- Air Flames with Complex Chemistry. *Combustion Science and Technology*, Vol. 60, pp. 267-284.

De Joannon, M., Saponoaro, A. & Cavaliere, A. (2000), Zero Dimensional Analysis of Diluted Oxidation of Methane in Rich Conditions, *Proceedings of the Combustion Institute*, Vol.28, pp. 1639-1646.

Eddings, E. & Sarofin, A. (2000), Advances in the use of Computer Simulations for Evaluating Combustion Alternatives, *3rd CREST Int. Symp.*, Yokohama, March 7-9, 2000.

Giovangigli, V. & Smooke, M.D. (1987), Extinction Limits of Strained Premixed Laminar Flames with Complex Chemistry. *Combustion Science and Technology*, Vol. 53, pp. 23-49.

Glassman, I. (1996), *Combustion*, 3rd Ed., Academic Press, New York.

Hamdi, M., Benticha, H. & Sassi, M. (2004), NO_x Emission from High Temperature Air versus Methane/Air Counter-flow Partially Premixed Flame, *IFRF Combustion Journal*, Article Number 200403, Retrieved from <http://www.journal.ifrf.net/>

Hamdi, M., Benticha, H., & Sassi, M. (2009), Numerical Modeling of the Effect of Fuel Dilution and Strain Rate on Reaction Zone Structure and NO_x Formation in

Flameless Combustion. *Combustion Science and Technology*, Vol. 181, No. 8, pp. 1078-1091.

Hasegawa, T., Tanaka, R. & Niioka, T. (1997), Combustion with High Temperature Low Oxygen Air in Regenerative Burners, *The first Asia-Pacific Conference on Combustion*, pp. 290-293, Osaka, Japan.

Ju, Y. & Niioka, T. (1997), Computation of NO_x Emission of a Methane–Air Diffusion Flame in a Two-Dimensional Laminar Jet with Detailed Chemistry. *Combust. Theo. Model.*, Vol. 1, pp. 243- 258.

Katsuki, M. & Hasegawa, T. (1998), The Science and Technology of Combustion in Highly Preheated Air Combustion, *Proceedings of the Combustion Institute*, Vol. 27, pp. 3135-3146.

Kee, R.J., Miller, J.A., Evans, G.H., & Dixon-Lewis, G. (1988). A Computational Model of the Structure of Strained, Opposed Flow, Premixed Methane-Air Flames. *Proceedings of the Combustion Institute*, Vol. 22, pp. 1479-1494.

Kee, R. J., Rupley, F. M., Miller, J. A., Meeks, E., Lutz, A. E., Warnatz, J. & Miller, S. F. (2002), PaSR Application User Manual: Modeling the Mixing and Kinetics in Partially Stirred Reactors, *Chemkin Collection Release 3.7*, Reaction Design, Inc., San Diego, CA.

Kim, S.H., Huh, K.Y. & Dally, B.B. (2005), Conditional Moment Closure Modeling of Turbulent non-Premixed Combustion in Diluted Hot Co-flow, *Proceeding of the Combustion Institute*, Vol. 30, pp. 751-757.

Kumar, S., Paul, P.J. & Mukunda, H.S. (2002), Studies on a new High-Intensity Low-Emission Burner, *Proceedings of the Combustion Institute*, Vol. 29, pp. 1131-1137.

Lutz, A.E., Kee, R.J., Grcar, J.F. & Rupley, F.M. (1996), OPPDIF: A Fortran Program for Computing Opposed-Flow Diffusion Flames. *Sandia National Laboratories Report*, SAND96-8243.

Mancini, M., Weber, R. & Bollettini, U. (2002), Predicting NO_x Emission of a Burner Operated in Flameless Oxidation Mode, *Proceedings of the Combustion Institute*, Vol. 29, pp. 1153-1163.

Milani, A. & Saponaro, A. (2001), Diluted Combustion Technologies. *IFRF Combustion Journal*, Article Number 200101,. Retrieved from <http://www.journal.ifrf.net/>

Miller, J.A., Kee, R.J., Rupley, F.M. & Meeks, E. (1996), CHEMKIN III: A Fortran Chemical Kinetics Package for the Analysis of Gas-Phase Chemical and Plasma Kinetics, *Sandia Report SAND 96-2816, Sandia National Laboratories*, Livermore, USA.

Muruta, K., Muso, K., Tkeda, K. & Niioka, T. (2000), Reaction Zone Structure in Flameless Combustion, *Proceedings of the Combustion Institute*, Vol. 28, pp. 2117-2123.

Nishimura, M., Suzuki; T., Nakanishi, R. & Kitamura, R. (1997), Low-NO_x Combustion under Highly Preheated Air Temperature Condition in an Industrial Furnace. *Energy Conversion and Management*, Vol. 38, pp. 1353-1363.

Özdemir, I. B. & Peters, N. (2001), Characteristics of the Reaction Zone in a Combustor Operating at MILD Combustion. *Experiments in Fluids*, Vol.30, pp. 638-695.

Plessing, T., Peters, N. & Wünning, J.G. (1998), Laser-optical Investigation of Highly Preheated Combustion with Strong Exhaust Gas Recirculation, *Proceedings of the Combustion Institute*, Vol. 27, pp. 3197-3204.

Pope, S. B. (1991), A Monte Carlo Method for the PDF Equations of Turbulent Reactive Flow. *Combustion Science and Technology*, Vol. 25, pp. 159-174.

Szegö, G.G, Dally, B.B., Nathan, G.J. & Christo, F.C. (2007), Performance Characteristics of a 20 kW Mild Combustion Furnace, *6th Asia-Pacific Conference on Combustion*, vol. 27, pp. 1157-1165.

Weber, R., Orsino, S., Lallement, N. & Verlaan, A. (2000), Combustion of Natural Gas with High-Temperature Air and Large Quantities of Flue Gas, *Proceedings of the Combustion Institute*, Vol. 28, pp. 1315-1321.

Weber, R., Verlaan, a.L., Rrsino, S. & Lallement, N. (1999), On Emerging Furnace Design Methodology that Provides Substantial Energy Savings and Drastic Reductions in CO_2, CO and NO_x Emissions. *Journal of the Institute of Energy*, Vol. 72, pp. 77-83.

Williams, F.A. (1985), *Combustion Theory*. 2nd Ed, The Benjamin/Cummings Publishing Company, Inc.

Wünning, J.A. & Wünning, J.G. (1997), Flameless Oxidation to Reduce Thermal NO-Formation. *Progress in Energy Combustion and Science*, Vol. 23, pp. 81-94.

Part 3

Functional Analysis and Health Monitoring

5

A New Expert System for Load Shedding in Oil & Gas Plants – A Practical Case Study

Ahmed Mahmoud Hegazy
Engineering for Petroleum and Process Industries (Enppi), Cairo
Egypt

1. Introduction

Electrical Power source of gas turbine driven generators is widely used in oil & gas industries.

It is the most adequate means due to availability of natural gas as a fuel to prime movers.

This type of power generation is in most cases used in isolation from the utility source as an island generation pool.

In critical operation conditions at which the generation drops severely , a case of insufficiency between generation and loading is established. In such a critical case a compromise between maintaining the process production and whole area power outage is made by shedding the non essential loads. Shedding of these loads is made till power is restored by the operators.

This chapter presents an integrated knowledge base system for generation dropping monitoring in oil& gas plants. The knowledge-based designed system monitors clearly and continuously the performance of power generation and distribution during operation. Effects of different generation insufficiency probabilities resulting from one or more of power turbine sudden shutdown are investigated. The developed system checks the stability of the system after generation dropping disturbance with power flow prior& after the disturbance and at each load shedding step to insure safe and stable operation of the industrial power system.

It indicates the system frequency and its deviation, voltage levels on established buses, generator machines operating point, real and reactive power flow.

The developed system discusses the overload capability of the generator as well. It indicates the recommended loads to be shed at discrete time events. It generates also system parameters and operating points curves.

The Main novelty of this approach is utilizing a built-in knowledge base to automatically remember the system configuration, operation conditions as load is added **or removed**. Thus predicting the system response to different disturbances. So it makes fast, correct and reliable decisions on load shedding priority. Obviously it is more suitable to industrial applications.

Whilst the modern utility load shedding applications use more monitoring and mining network data streams is crucial for managing and operating data networks. Such systems must cope with the effects of overload situations due to large volumes , high data rates and bursty nature of the network traffic. Some Intelligent trials use adaptive load shedding strategy by executing the artificial neural network (ANN) and transient stability analysis for an Industrial cogeneration facility. To prepare the training data set for ANN, the transient stability analysis has been performed to solve the minimum load shedding for various operation scenarios without causing tripping problem of cogeneration units.

2. The developed system

Large industrial plants usually feed large electric loads at different voltage levels.

Severe losses are foreseen in case of loss of power supply. Accordingly, the load shedding scheme to control the process of generation drop to counterpoise the most needed loads so as to maintain production till the power supply problem clearance, is a real serious vital job.

Most commercial load shedding schemes are using independent system under-frequency relays that detect the frequency drop limit and start to trip loads in stages according to a predefined truth table. Other alternatives are using the SCADA software on power management system to detect the number of operating power generators to decide the start of load shedding.

Power system stability is the property of a power system that insures that the system remains in electromechanical equilibrium throughout any normal and abnormal conditions (i.e. generator operating point remains in the limits indicated on manufacturer power chart). One of major causes of industrial power system instability problems is generation drop (unbalance between generation and load demand). The consequences of instability problems are very severe and may cause permanent damage on equipment, moreover the whole area power outage. Stability studies are dynamic simulations of the particular groups of machines in the system that are known to have important influence on the system operation, in order to study system frequency variations , power output,...etc. One of tools for power system enhancement is adding load shedding scheme.

A stability study is performed to investigate dynamic conditions on the power system during generation. The study shall begin with steady state load flow conditions for the purpose of establishing power flows (watt & Var) in the system at time intervals immediately prior to the system disturbances under investigation. The dynamic studies shall determine the following: machine operating point, real & reactive power flows for selected machines, system frequency & frequency deviation, voltage levels on established buses at disturbance initiation, discrete times during and following its removal from the system. The conditions for each time interval shall be printed out and the values for each time interval shall be used as the initial condition for the subsequent interval during the entire period studied to obtain a definite stability conclusion.

In this chapter a new method for load shedding is presented using the transient stability studies to analyze this problem. It is deemed necessary to make dynamic simulations of the particular group of machines in the system that have major influence on system operation.

A combined system for monitoring of generation and load demand to load shedding in case of unbalance between them has been developed. An integrated software has been designed to perform this vital task.

The developed system is composed of three different modules; input and general studies modules are constructed and carried out supported by the Electrical Transient Analyzer Program (ETAP). Outputs of these modules are automatically collected and indexed within the knowledge base module. The knowledge base module contains mainly the generator power chart constraints, generator thermal limitations and load shedding design rules. Also it contains switching events rules as well as a list of load categories recommended by process engineers. The knowledge base and output modules are developed using Matlab Simulink. As well a graphic module has been developed to visualize the system and record generator frequency, generator exciter voltage, generator output power and generator current. The developed system is designed in a modular form and consists mainly of three modules as described in the following sections.

2.1 Input module

This module constructs the data base which represents the plant. The input module specifies the plant description, edits the single line diagram and defines the operation of different loads using colour code. Using this module, the power plant will be fully described and the following data is transferred to the knowledge base module automatically:

a.   Plant bus-bars connection.
b.   Location of each load in the plant map.
c.   Feeders and cables data. A special library is used for cable description.
d.   Load description, specification, category and duty.

The load Category is classified to serve the load shedding priority as category (0) stands for important loads that can not be shed, category (1) stands for second priority loads to be shed with minimum impact on process for load shedding duration, category (2) stands for first priority loads to be shed with no process impact for load shedding duration.

The system also accepts different definitions of power consumption for each load as follows:

a.   Design active power is defined as the active power for a driven machine or load in its normal mode of operation,
b.   Design particular power is defined as the active power for a driven machine or load in a load in a special loading conditions,
c.   Nameplate power is a rated active power as stated in the nameplate of a driven machine or load.

2.2 General studies module

This module is responsible for carrying out all plant calculations such as load flow study and transient stability study.

2.2.1 Load flow study

The load flow calculation tool specifies the plant operating conditions. It generates branch current loading, branch losses, bus loading and transformer loading. It generates also alarms for critical equipment and critical buses. For monitoring of motors operation in large plants, the load flow runs are made based on possible load and their results are considered as initial operating conditions. These runs are then repeated for load shedding steps to check the power system operating conditions at these circumstances.

2.2.2 Transient stability study

This is performed to investigate dynamic conditions on power system during generation drop. The dynamic simulations of a particular group of machines in the system that are known to have important influence on the system operation is performed. The aim of this study is to calculate system frequency variations, power output...etc. Fig.(1) indicates the AVR/exciter model applied for the case study presented in this chapter.

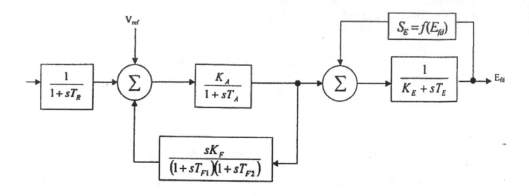

Fig. 1. AVR/Exciter model

2.3 Knowledge base module

This module is the heart of the system. It consists of two types of the rule base that manipulates the load shedding steps. Type (I) is a time event of generation dropping and load shedding actions in the plant. Type(II) is concerned with generator stability check.

This expert knowledge is normally stored separately from the procedural part of the program. It is a form of data base containing both facts which describe fixed properties or characteristics of generation & loads and information on how to do load shedding.

As an example the statement:

"A3 is a circuit breaker" is a fact.

If it is desired to shed network by opening a circuit breaker, a rule could be formed expressing such knowledge in the form of IF (X) THEN (Y), where X is the premise or antecedent and Y is the conclusion or consequent, which is either True or false.

2.3.1 Type I

The rule base specifies the process constraints and the events expected in each plant. This part of the rule base has a storage capacity of time events from (t1 to t15). Each time event allows 10 actions to be defined. Fig.(2) shows a time event page.

2.3.2 Type II

This type of rule base identifies the level of control and monitoring for generation and load demand as well as recommended settings. It may include up to 300 rules for a similar number of motors. It classifies the motors based on motor type, motor ratings and operation criteria as well. The following constraints are some rules in the rule base:

a. The generator thermal withstand monitoring unit: This part covers many items such as generator manufacturers thermal withstand characteristics as related to the fault current and duration that the machine can withstand. This unit is of prime importance to check continuously the generator loading to decide the instant of load shedding initiation.

   The load category unit: This unit is for the load category as classified to serve the load shedding priority category (0,1,2).

b. Generator power limiter unit: This unit supervises the generators power capability with respect to many limits such as rated stator current, rotor current limit, practical stability limit, minimum excitation, power factor, lead & lag domains...etc.

c. Generator power angle check unit: This unit provides the check for power angle as a stability index for operating generator.

d. Load shedding control unit: This unit determines the number of loads, the steps, and the duration of each step to apply a proper load shedding scheme.

The execution is based on the knowledge base contains six rules of the following form:

IF (A) THEN (B)

Where:

A: is the condition part of the rule.

B: is the action part of the rule.

Each Rule corresponds to a path in the decision tree in Fig(3) that provides an established technique for solving classification problems, which have a small number of categories stable versus unstable.

The set of rules are as follows:

Rule 1: IF the power angle of each generator in the system with respect to fixed reference in the system increases to a maximum value and then decreases THEN the system is considered as synchronously stable no load shedding.

Rule 2: IF the system is synchronously stable,

   AND the final voltage profile of the system is acceptable.

   AND the final loading of every branch in the system is acceptable.

   THEN the final system condition is considered as acceptable.

   AND the run can be terminated.

Rule 3: IF the system is synchronously stable,

   AND the final voltage profile of the system is unacceptable due to violation of the voltage limit in one more of the system bus THEN load shedding is needed to correct the final voltage profile.

Rule 4: IF the system is synchronously stable,

   AND the final loading profile of the system is unacceptable due to violation of the loading limit in one or more of the system branches THEN load shedding is needed to correct the final loading profile.

Rule 5: IF the power angle of one or more generators in the system with respect to fixed reference in the system increases indefinitely THEN the system is considered as synchronously unstable.

Rule 6: IF the loading is higher than generation.

AND the system is synchronously unstable.

AND the power angle of certain generator (i) increases indefinitely.

THEN load shedding is performed as a control action.

The rules from (1) to (6) have to be repeated till a stable position is reached.

The interpreter tool that links between ETAP and Matlab Simulink is the UDM control blocks. The UDM control blocks are created using Matlab Simulink interface tool.

The components of the new expert system are shown in Fig.(4). The four major components of this system are the user interface , the simulation program , the rules for transient stability evaluation & for suggesting load shedding control actions and the interface between them. The operation logic is described in the flow chart Fig.(5).

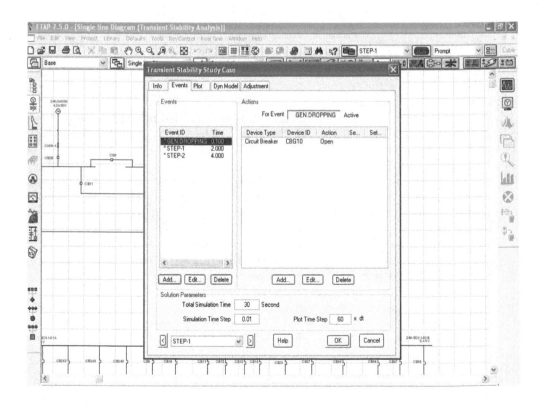

Fig. 2. Recommended Time Event unit for generator dropping and load shedding actions.

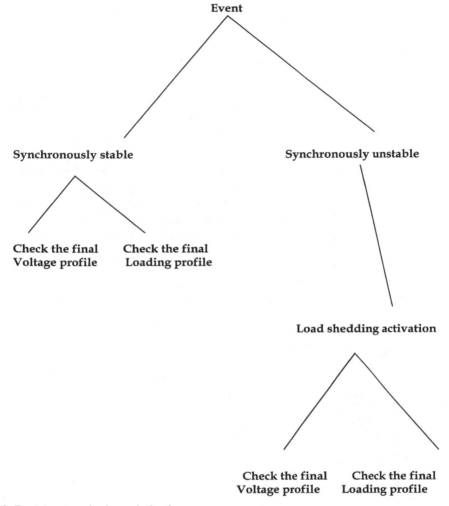

Fig. 3. Decision tree for knowledge base.

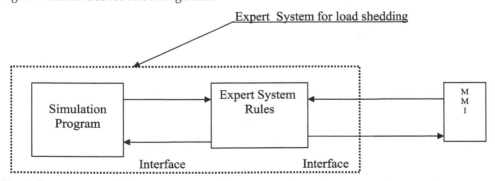

Fig. 4. Components of the new expert system.

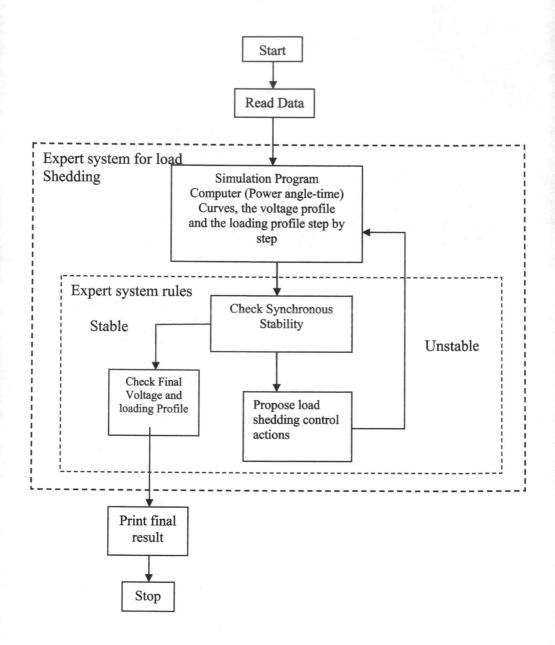

Fig. 5. The operation logic flow chart of the new expert system.

3. The case study

The industrial power system investigated here is a typical natural gas processing plant onshore facility. The loads of the plant are supplied at two voltage levels. The ratings range from 1MVA motors down to fractions of horsepower rating loads. The power supply to the plant consists of three gas turbine driven generators each rated 4.24MW at 0.8 power factor. Two gas turbines are continuously feeding the demand loads and the third one is in standby mode.

For clarification, the simplified single line diagram is shown in Fig.(6).

During severe conditions of the power generation system one turbine is carrying the total plant load which results in insufficient generation to meet loads , especially in Summer time where the turbine rating is de-rated due to high ambient temperature.

The condition in this abnormal operation , the second turbine will be able to share the load with the running turbine within one hour. After the duration is elapsed the plant must be shutdown in case the second turbine is failed to start. Automatic load shedding program preset to drop specific loads to steer the power system operation into safe margins.

The load shedding system shall be implemented in the substation switchgear by having initiation signal from turbine emergency system (voltage free contacts from each turbine to indicate turbine shutdown) to the substation supervisory system. The speed of communication is a dominating factor to achieve a rapid load shedding system within the overload capacity of the generator.

These hardwired signals are processed in the substation supervisory system which would operate the load shedding scheme the instant it detects that 2 of the 3 turbines are stopped (off) and the third is supplying the whole load. The status of the generator circuit breakers are used as a confirmation signal. The consumed power measurements on each generator circuit breaker monitored by the substation supervisory system shall be used to determine that the only running generator loading above 3.7MW.

The loads in switchgear panel tag item 240-ESI1-02 would be tripped through the already wired stop signals (inhibit to start for the load shedding duration) and a signal would be transmitted via fibre optic cable to trip the loads (inhibit to start for the load shedding duration) in panel 240-ESI1-02.

3.1 Input and general studies output

A load flow study had been done for the system under consideration. Table (1) shows the results of this study for normal operation of the system in which two turbines carry the demand loads.

As generation drop of one generator causes the system to no longer operate in safe margins and therefore loads have to be shed to remove the overload on the still operating turbine-generator.

The system parameters are shown in the following curves are depicted versus time

Fig.(7) speed response at generation drop.

Fig.(8) Generator electrical power demand.

Fig.(9) Generator terminal current.

Fig.(10) Generator exciter voltage.

Fig.(11) Generator exciter current.

In critical operating conditions the load shedding steps are used to remove the only running turbine overload. Item 3.2 indicates the results of applying load shedding steps. The generator thermal withstand curve indicates that the machine can sustain 3 times nameplate current for 10 seconds and overload capability is 120% for 5 seconds. Accordingly the amount of load shed at every stage should keep the generator overload less than 120%. After application as described in item 3.2, the system parameters after load shedding are shown in the following curves are depicted versus time:

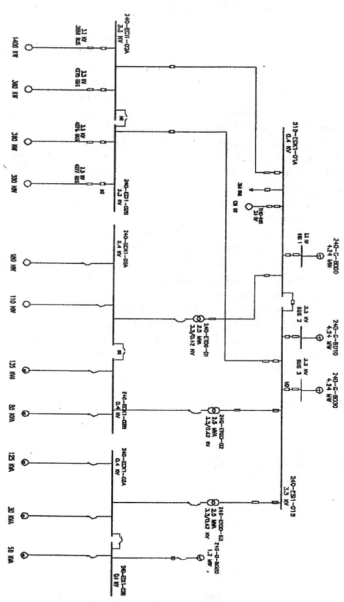

Fig. 6. Global Single Line Diagram.

Bus			Generation		Motor load		Static load		Load flow					
ID	KV	%MAG	ANG	MW	MVAR	MW	MVAR	MW	MVAR	ID	MW	MVAR	AMP	%PF	%TAP
240-ECK1-01A	0.4	102.714	-1.4	0	0	0.33	0.23	0.38	0.29	ECK1-01A	-0.71	-0.52	123.5	91.7	
240-ECK1-01B	0.4	103.314	-1	0	0	0.12	0.09	0.39	0.29	ECK1-01B	-0.51	-0.39	897.2	80	
240-ECK1-02A	0.4	100.279	-2.7	0	0	0.09	0.07	0.43	0.47	ECK1-02A	-1.17	-0.98	210.3	80	
										240-ECK1-02B	0.45	0.34	803.7	80	
240-ECK1-02B	0.4	100.279	-2.7	0	0	0.01	0.01	0.44	0.33	240-ECK1-02A	-0.45	-0.34	803.7	80	
240-EC11-01A	3.3	99.978	0	0	0	0	0	0.28	0.21	BUS 1	-2.88	-1.72	587.6	85.8	
										7100 - BUS	0.07	0.06	16.8	73.9	
										ECK1-01A	0.71	0.55	157.1	79.3	
										ES11-02A	2.64	0.95	492	94.1	
										240-ES11-01B	-0.83	-0.05	144.8	99.8	
240-ES11-01B	3.3	99.978	0	0	0	0	0	0	0	BUS2	-3.31	-1.98	674.3	85.8	
										ECK1-0-1B	0.52	0.4	114.2	79	
										ECK1-02A	1.18	0.97	267.7	77.1	
										ES11-02B	0.79	0.55	168.1	81.9	
										240-ES11-01A	0.83	0.05	144.8	99.8	
7100-BUS	3.3	99.954	0	0	0	0.07	0.06	0	0	240-ES1-01A	-0.07	-0.06	168	73.9	
BUS 1	3.3	100	0	2.98	1.72	0	0	0	0	240-ES11-01A	2.88	1.72	587.5	85.8	
BUS 2	3.3	100	0	3.31	1.98	0	0	0	0	240-ES11-01B	3.31	1.98	674.2	85.8	
BUS 7	15	103.524	-1.6	0	0	0	0	0	0	240-ES12-01	0	0	0	0	
BUS 9	3.3	98.914	-0.1	0	0	1.37	0.9	0	0	3550BUS	-1.37	-0.9	291.1	83.5	
ECK1-01A	3.3	98.961	0	0	0	0	0	0	0	240- ES11-01A	-0.71	-0.55	157.1	79.3	
										240-ECK1-1A	0.71	0.55	157.1	79.3	
ECK1-01B	3.3	99.963	0	0	0	0	0	0	0	240-ES11-01B	-0.52	-0.4	114.2	79	
										240-ECK1-01B	0.52	0.4	267.7	79	
ECK1-02A	3.3	99.883	0	0	0	0	0	0	0	240-ES11-01B	-1.18	-0.97	2667.7	77.1	
										240-ECK1-02A	1.18	0.97	367.7	77.1	
ES11-02A	3.3	99.498	-0.1	0	0	0	0	0	0	240-ES11-01A	-2.63	-0.95	492.3	94.1	
										240-ES11-02A	2.63	0.95	492.3	94.1	
ES11-02B	3.3	99.801	0	0	0	0	0	0	0	240-ES11-01B	-0.79	-0.56	168.6	81.6	
										240-ES11-02B	0.79	0.56	168.6	81.6	

Table 1. Global load flow report for system normal operation

Fig.(12) speed response at generation drop after load shedding.
Fig.(13) Generator electrical power demand after load shedding.
Fig.(14) Generator terminal current after load shedding.
Fig.(15) Generator exciter voltage after load shedding.
Fig.(16) Generator exciter current after load shedding
Fig.(7) shows that the generator speed drops 1465 rpm, then it rises and keeps steady at 1475 rpm (approximately keeps fixed at -1.6% drop from rated speed).

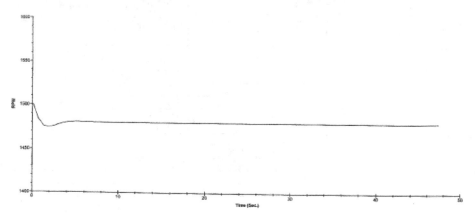

Fig. 7. Speed response at generation drop.

Fig.(8) shows that the generator electrical power demand rises to 4.5 MW to reflect generator overload capability to supply the system requirements after generation drop.

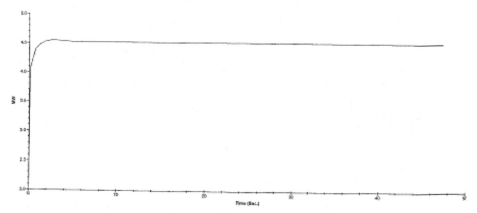

Fig. 8. Generator electrical power demand.

Fig.(9) shows that the generator terminal current rises to 950A to reflect generator overload capability to supply the system requirements after generation drop.
Fig.(10) shows that the generator exciter voltage rises to 3.25 p.u. representing the behaviour of the regulated exciter to accommodate the overload demand.
Fig.(11) shows that the generator exciter current rises to 3.25 p.u. representing the behaviour of the regulated exciter to accommodate the overload demand.

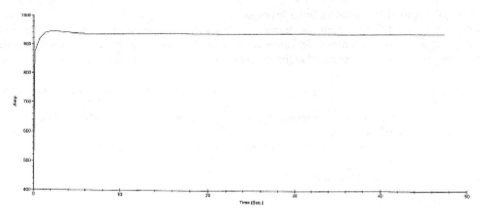

Fig. 9. Generator terminal current.

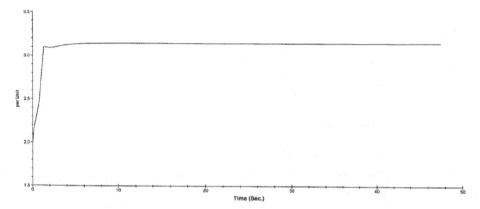

Fig. 10. Generator exciter voltage.

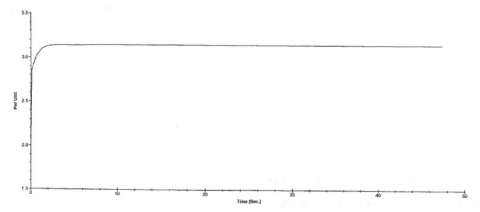

Fig. 11. Generator exciter current.

3.2 Application of knowledge base module

Applying this module, the plant conditions in each stage during generation drop and operation is investigated and printed out. By applying this module also, the module has selected the following load shedding steps so that the power system can operate in safe margins:

- Step 1. Flash gas compressor rated 1400kW (consumed power 1375kW) at 2 seconds,
- Step 2. Condensate export pumps each is rated 360kW (consumed power 160kW each) at three seconds,
- Step 3. Residential camp consumed active power of 280kW, and 470kW low voltage motor loads at 4 seconds.

The system new operating conditions after the execution of the load shedding 3 steps are shown in the following figures:

Fig.(12) shows that the generator speed rises up to 1500 RPM.

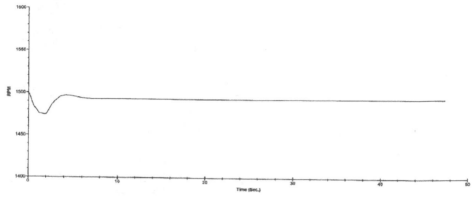

Fig. 12. Speed response after load shedding

Fig.(13) shows that the generator electrical power demand falls to 3 MW after the three steps which is machine safe limits.

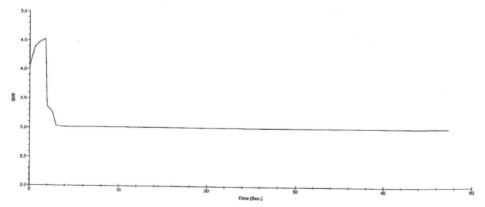

Fig. 13. Generator electrical power demand after load shedding.

Fig.(14) shows that the generator terminal current falls to 640A to reflect generator overload capability to supply the system requirements after load shedding.

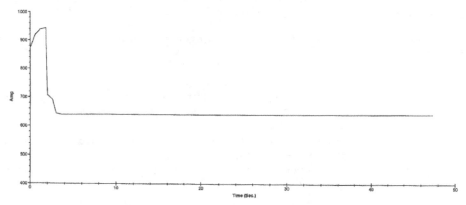

Fig. 14. Generator terminal current after load shedding.

Fig.(15) shows that the generator exciter voltage drops to 2.4 p.u. representing the behaviour of exciter after load shedding.

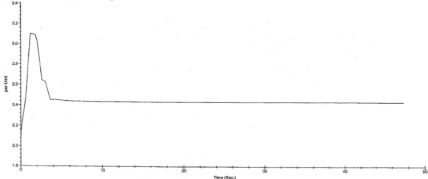

Fig. 15. Generator exciter voltage after load shedding.

Fig.(16) shows that the generator exciter current rises to 2.4 p.u. representing the behaviour of the exciter after load shedding.

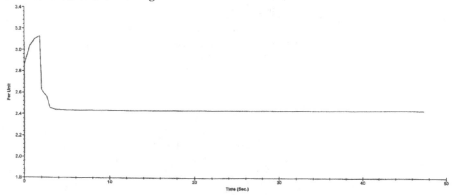

Fig. 16. Generator exciter current after load shedding

4. Conclusion

In this chapter, a new expert system for load shedding in large industrial plant is developed and has been presented.

This expert system is better than the traditional stand alone under frequency relays that sense the frequency and trip under pre-defined values which are not responsive dynamically to the system, whereas the new expert system presented in this chapter is a rule based system for generation monitoring and load shedding which is flexible with the system dynamics. It has been developed in MATLAB software.

This system is supported by ETAP software package for power system analysis to act as comprehensive tool for power system operation monitoring in large industrial plant. The system is tested and the results obtained for a study system are presented to assure reliable performance of the developed system.

5. References

B. Babcock, M. Datar and R. Motwani. *Load shedding techniques for data stream systems,2003*.

B. Babcock, M. Datar and R. Motwani. *Load shedding techniques for aggregation queries over data streams*. In ICDE,2004.

Barzan Mozafari and Carlo Zaniolo. Optimal Load Shedding with Aggregates and Mining Queries. 26th IEEE International Conference on Data Engineering, March 1-6, 2010, Long Beach, California: 76-88.

Cheng-Ting Hsu; Hui-Jen Chuang; Chao-Shun Chen. *Artificial Neural Network Based Adaptive Load Shedding for an Industrial Cogeneration Facility*. Industry Applications Society Annual Meeting, 2008. IAS '08. IEEE .

Electrical Transient Analyzer Program ETAP. *Power station 4.0 Analysis module*, Operation Technology, Inc., February 2002.

Electrical Transient Analyzer Program ETAP. *Power station 4.0 user guide*, Operation Technology, Inc., February 2002.

Horowitz, S.H.; Politis, A.& Gabrielle, A.F. (1971). *Frequency Actuated Load Shedding and Restoration part II-implementation*, IEEE trans. Power App. Syst. Steven Rovnyak (1994). Decision trees for real-time transient stability prediction. IEEE trans. On Power System Vol.9 No.3.

IEEE STD.399, IEEE Recommended practice for industrial and commercial power system analysis.

Malizewski, R.M.; Dunlop, R.D. & Wilson, G.L. (1971). *Frequency Actuated Load Shedding and Restoration part I-philosophy*, IEEE trans. Power App. Syst.

Steven Rovnyak (1995). Predicting future behaviour of transient events rapidly enough to evaluate remedial control options in real-time. IEEE trans. On Power System Vol.10 No.3.

Tatbul,N.,ET AL. *Load shedding in data stream manager*. In Proceedings of intl.conf. on very large data bases 2003,pp.309-320.

Yasser, G.M. (2004). A Knowledge Based Integrated Motor Protection and Monitoring System for Large Industrial Plant , *Proceedings of Scientific Bulletin of Faculty of Engineering Ain Shams University*, vol.39 # 1, ISBN 842-6508-23-3, Cairo, Egypt, March 31, 2004.

Yijian Bai, Haixun Wang, Carlo Zaniolo. *Load shedding in classifying multi-source streaming data: A Bayes risk approach*. Technical report TR060027,UCLA CS Department 2006.

Yun Chi, Philip S. Yu, Haxiun Wang and Richard Muntz. *Loadstar: A load shedding scheme for classifying data streams*. In Proceedings of SIAM intl. conf. on data mining 2005.

6

Application of Functional Analysis Techniques and Supervision of Thermal Power Plants

M.N. Lakhoua
Member IEEE
Laboratory of Analysis and Command of Systems (LACS)
ENIT, Le Belvedere, Tunis,
Tunisia

1. Introduction

Supervision consists of commanding a process and supervising its working [1]. To achieve this goal, the supervisory system of a process must collect, supervise and record important sources of data linked to the process, to detect the possible loss of functions and alert the human operator.

The main objective of a supervisory system is to give the means to the human operator to control and to command a highly automated process [2]. So, the supervision of industrial processes includes a set of tasks aimed at controlling a process and supervising its operation.

An automatic supervisory system is a traditional supervisory system, that is to say, a system which provides a hierarchical list of alarms generated by a simple comparison with regard to thresholds [3]. The information synthesis system manages the presentation of information via any support (synoptic, console, panel, etc.) to the human operator.

Today, the supervision of production systems is more and more complex to perform, not only because of the number of variables always more numerous to monitor but also because of the numerous interrelations existing between them, very difficult to interpret when the process is highly automated [4].

The challenge of the future years is based on the design of support systems which let an active part to the supervisory operators by supplying tools and information allowing them to understand the running of production equipment. Indeed, the traditional supervisory systems present many already known problems. First, whereas sometimes the operator is saturated by an information overload, some other times the information under load does not permit them to update their mental model of the supervised process [5].

Moreover, the supervisory operator has a tendency to wait for the alarm to act, instead of trying to foresee or anticipate abnormal states of the system. So, to avoid these perverse effects and to make operator's work more active, the design of future supervisory systems has to be human centred in order to optimize Man-Machine interactions [6].

It seems in fact important to supply the means to this operator to perform his own evaluation of the process state. To reach this objective, Functional Analysis seems to be a promising research method. In fact, allowing the running of the production equipment to be understood, these techniques permit designers to determine the good information to display through the supervisory interfaces dedicated to each kind of supervisory task (monitoring, diagnosis, action, etc.). In addition, Functional Analysis techniques could be a good help to design support systems such as alarm filtering systems [7].

By means of a significant example, the objective of this paper is to show interests of the use of Functional Analysis (FA) techniques such as SADT (Structured Analysis Design Techniques) and SA-RT (Structured Analysis Real Time) for the design of supervisory systems. An example of a SCADA system of a thermal power plant (TPP) is presented. The next section briefly describes the characteristics of a SCADA system and the problems linked to its design. Next, the interests of using FA in the design steps are developed. In Section 3, after presenting concepts of SADT and SA-RT, these methods are applied to a SCADA system of a TPP. The last section presents a discussion about the advantages and inconveniences of FA techniques and some examples of SCADA applications in a TPP.

2. Presentation of supervisory control and data acquisition system

The SCADA term (supervisory control and data acquisition) refers to a system that collects data coming from different sensors of an industrial or other process [8] [9] [10], these sensors can beings installed in the same site or distant (several Km), the introverted data are treated by an unit called processor power station (CPU, PCU, PC...), results are sent in real time to the Men / Machine interfacing that can be a computer with its peripherals (See Figure 1).

The SCADA system of the TPP of Radès (in Tunisia) orders and classifies all data for [11] [12]:

- Instantaneous impression.
- Visualization on screen using data tables and tabular diagrams.
- Registration of instantaneous exchanges of numeric and analogical data.
- Instantaneous calculation for example corrections of gas debits, direct middle specific consumption, middle values.
- Storage of the analogical information of the process.
- Calculation of outputs and losses of the process.
- Surveillance of the SOE signals (entrances rapid contact 1ms)
- Interfacing interactive Men / Machine for the surveillance of the system and the conduct of processes (tabular, curves view of alarm).

The objective of the SCADA system of the TPP of Radès is to collect data instantaneously (ON LINE) of their sites and to transform them in numeric data. This centralized supervision allows operators, since the room of control of the TPP, to control facilities in their domain of exploitation and the different types of incidents.

The SCADA system of the TPP of Radès is equipped of three networks of communication:

- Field bus, 5 Mbits, permitting to do exchanges of the numeric data of the entrance card / exits (FBM) toward the central system (CP) via modules of communication (FCM);
- Node bus, 10 Mbits, permitting to do exchanges of the numeric data of the central system (CP) via modules of communication (DNBT) toward the Men/Machine interfacing (workstations);

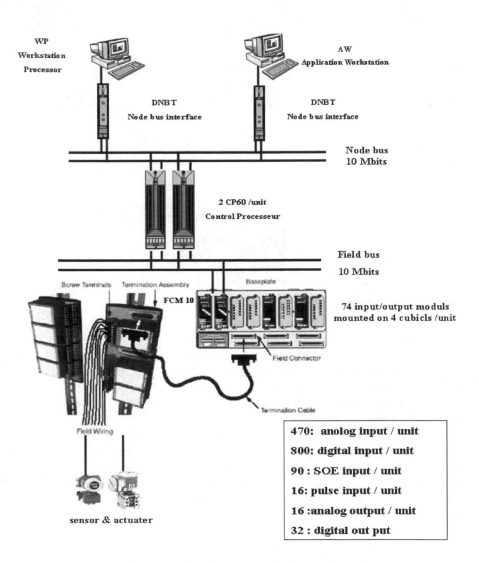

Fig. 1. SCADA system of the TPP.

- Ethernet TCP/IP, 100 Mbits, permitting to do exchanges of files between workstations of the Men/Machine interfacing. It avoids so the overcharge of the Nodebus network.

Figure 2 presents the different links between the CP60, FCM and FBM blocks.

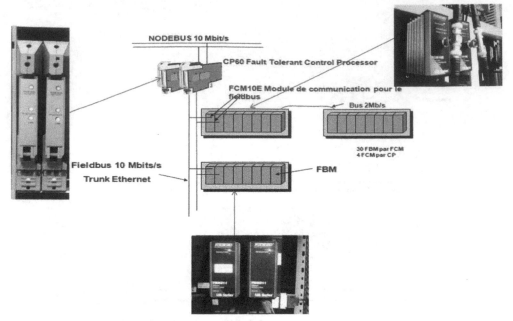

Fig. 2. The CP60/FCM/ FBM links.

3. Functional analysis of a SCADA system

There are many methods that have been used to enhance participation in Information System planning and requirements analysis [13]. We present the application of SADT and SA-RT methods on a SCADA system.

The SADT method represents attempts to apply the concept of focus groups specifically to information systems planning, eliciting data from groups of stakeholders or organizational teams. SADT is characterized by the use of predetermined roles for group/team members and the use of graphically structured diagrams. It enables capturing of proposed system's functions and data flows among the functions [14].

SADT, which was designed by Ross in the 1970s [15], was originally destined for software engineering but rapidly other areas of application were found, such as aeronautic, production management, etc.

For SADT diagrams or function boxes, we will consider two events to be representing the activation states of the activities. The first event represents the instant when the activity is triggered off, and the second event represents the ending instant.

Figure 3 presents the Actigram and Datagram of the SADT model.

The boxes called ICOM's input-control-output-mechanisms are hierarchically decomposed. At the top of the hierarchy, the overall purpose of the system is shown, which is then decomposed into components-subactivities. The decomposition process continues until there is sufficient detail to serve the purpose of the model builder.

Figures 4 and 5 present respectively the A-0 and the A0 levels of the SADT model of a SCADA system [16] [17].

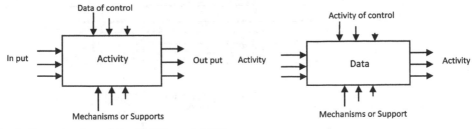

Fig. 3. Organization of the SADT model [19].

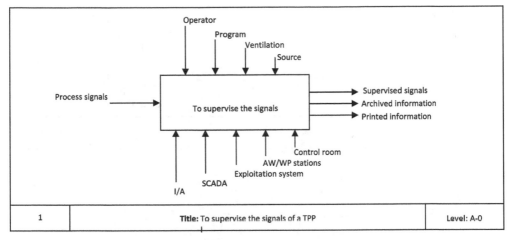

Fig. 4. A-0 level of the SADT model [16].

SADT is a standard tool used in designing computer integrated manufacturing systems, including flexible manufacturing systems. Although SADT does not need any specific supporting tools, several computer programs implementing SADT methodology have been developed. One of them is Design: IDEF, which implements IDEF0 method. SADT: IDEF0 represents activity oriented modeling approach.

Among the graphical methods most commonly used in industry, two of the leading methods are SA-RT and Statecharts. SA-RT is a short name for Structured Analysis Methods with extensions for Real Time [18]. The model is represented as a hierarchical set of diagrams that includes data and control transformations (processes). Control transformations are specified using State Transition diagrams, and events are represented using Control Flows. The other graphical and state based paradigm for specification of real time systems is Statecharts. The system is represented as a set of hierarchical states instead of processes. Each state can be decomposed into sub states and so on. The statecharts notation is more compact than the SA-RT notation and has been formally defined.

Structured Analysis for Real-Time Systems, or SA-RT, is a graphical design notation focusing on analyzing the functional behavior of and information flow through a system [19]. SA-RT, which in turn is a refinement of the structural analysis methods originally introduced by Douglass Ross and popularized by Tom DeMarco in the seventies, was first introduced by Ward and Mellor in 1985 and has thereafter been refined and modified by other researchers, one well-known example being the Hatley and Pirbhai proposal (See Figure 6).

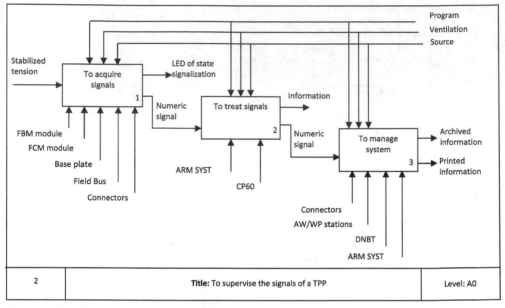

Fig. 5. A0 level of the SADT model [16].

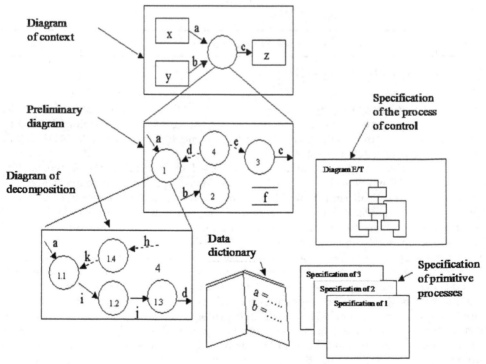

Fig. 6. Organization of an SA-RT model [19].

Thus, SA-RT is a complex method for system analysis and design. This is one of the most frequently used design method in technical and real-time oriented applications adopted by various Case-Tools. It is a graphical, hierarchical and implementation independent method for top-down development.

SA-RT method enables us to identify an entrance and an exit of data in an algorithm or a computer program. It is divided in three modules: Diagram of Context, Data Flows Diagram and Control Flows Diagram. Every module includes in its graphic interpretation different symbols.

So, the SA-RT model is composed exclusively of diagrams [20]. It starts with the main process 'To supervise the signals of the TPP' (Figure 7). Then, this process is broken into a preliminary data flows diagram composed of three processes. We continue the decomposition of the processes until the last decomposition level has been reached (levels DFD1, DFD2 and DFD3).

For this example, we present the control flow diagram of the application of the SCADA system in a hydrogen station (Figure 8).

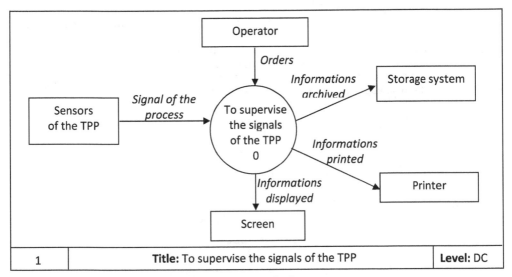

Fig. 7. Diagram of Context of the SCADA system [20].

In order to give a detailed vision of the control of the hydrogen station, we present on Figure 9 an example of a State/Transition diagram.

Compared to the results given by the SADT method, the SA-RT method allows a functional as well as a temporal analysis.

The possible uses for the SADT and the SA-RT models are the design of a monitoring display and a diagnosis display. For the design of a monitoring display, the A0 level of the SADT model or the preliminary data flow diagram of the SA-RT model supplies a global view of the system. Indeed, information relative to each function represented through this level should appear in a monitoring display.

For the design of hierarchical diagnosis displays, each actigram of the SADT model or data flows diagram of the SA-RT model constitutes a vision at a given abstraction level. So, each of these actigrams gives a less or more detailed vision. In function of the objectives defined by the designer for each display, a particular actigram or data flows diagram can supply the required information.

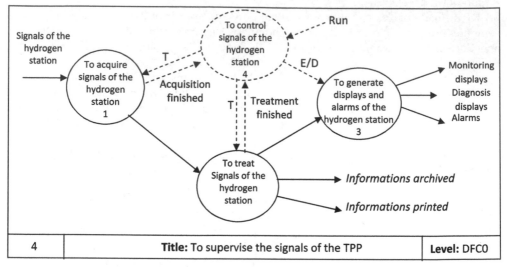

Fig. 8. Control Flows Diagram of the SCADA system [20].

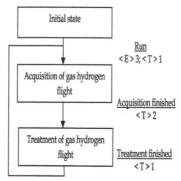

Fig. 9. State/Transition diagram of the SCADA system [20].

Finally, this application of the FA techniques on the SCADA system of a TPP shows briefly the interests of the SADT and SA-RT methods in the design of supervisory systems.

4. Application of the supervision of a hydrogen station of a TPP

The objective of this application is to show interests of the use of a SCADA system in a hydrogen station of a TPP. In fact, the cooling by hydrogen has been adopted for turbo-alternators in 1926. This technique has been used for the interior cooling of drivers while doing circulating of the fluid in their conducts, putting the fluid in contact with materials in which the heat is produced.

Hydrogen is an odorless, colorless, very light gas (more that air) and composed of two atoms of hydrogen. It possesses a high gravimetric energizing power: 120 MJ/kgs compared to oil (45 MJ/kgs), to the methanol (20 MJ/kgs) and to the natural gas (50 MJ/kgs). However it is as the lightest gas (2,016g/mol H_2), of where a weak volumetric power: 10,8MJ/m^3 facing the methanol (16 MJ/m^3), natural gas (39.77 MJ/m^3). It puts a real problem of storage and transport: that it is for the utilization of hydrogen in a vehicle or for the transport in pipeline, in truck, it is the volumetric density that imports. The volumetric energizing density of H2 is not interesting that to the state liquid tablet either (700 bars).

The fact that a mixture of hydrogen and air is an exploding mixture on a large range of proportions, the machine and the procedure of utilization specified is conceived so that no exploding mixture can occur in the normal conditions of working.

The aforesaid unforeseen conditions cannot present themselves that during the replacement. However the pressure of gas being nearly equal to the air pressure in this condition, the intensity of explosion can take place doesn't pass to the more of 7bar.

The purity of hydrogen H_2 in an alternator is always maintained superior to 95% until 98% and when it decreases to 90% an alarm is given out to the panel of the local cupboard as well as the room of control. Preventing gas H_2 intern to form an exploding the mixture. It is necessary to renew a certain volume of H_2 lodged in the alternator by another volume coming from bottles H_2.

The principle of the interior cooling permitted the increase of the strength of the alternator and an efficient utilization of the hydrogen pressure (See Figure 10).

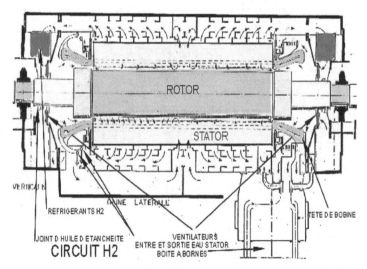

Fig. 10. Cooling by hydrogen of an alternator.

Gas hydrogen is introduced in the alternator while manipulating the regulator of pressure or the regulator. When the purity of hydrogen measured is 95% or more on the meter, its feeding is stopped, the regulator of pressure is adjusted foreseen at the level and the pressure of the alternator is increased. Thus, gas hydrogen is introduced in the envelope, giving back the ready alternator to the working. The diagram of circuit of the system of gas control represents the position of every floodgate during the working.

Otherwise, the drier of gas, composed of a full reservoir of alumina activated (absorbing agent), of a heating device, of a puff, of a thermometer…, is installed between the circuit high pressure and the circuit bass pressure of the alternator so that gas crosses the drier all along the working of the alternator.

The turbo-alternator group of the center of Radès is cooled internally by gas hydrogen. As shows the diagram of circuit (See Figure 11), the device of gas control is composed by the following elements:

- A spray of carbon dioxide
- A device of gas hydrogen feeding
- A drier of gas
- A unit of surveillance of gas pressure / purity
- Command valves
- A meter of purity.

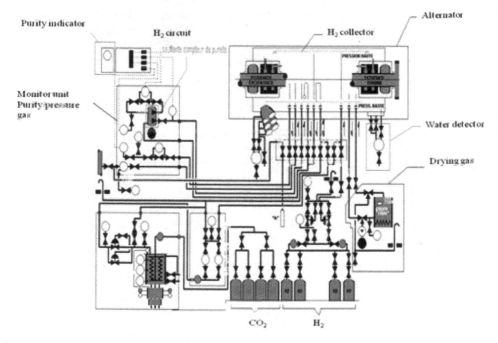

Fig. 11. Circuit of the gas hydrogen of the TPP.

The pressure of gas hydrogen inside the alternator is maintained to a face value of 3 to ABS 6 bars, thanks to a regulator of pressure gone up on the collector of feeding in hydrogen.

In the same way, the purity of hydrogen in the alternator is always maintained to more of 95% and, when it descends to 90%, an alarm is given out, preventing the internal gas to compose an exploding mixture. The gas of carbon dioxide is used fluid how of sweep to fill to either hunt the hydrogen of the alternator in order to avoid that hydrogen and air won't be mixed in a critical condition.

At the time of replenishment of the alternator by hydrogen, the dioxide of carbon is used to hunt the air of the alternator. The valve of safety is adjusted to an ABS 6 bar pressure so that, when an anomaly occurs in the circuit of gas of carbon dioxide, the pressure of the bottle is exercised on all tubings.

The CO_2 being heavier than air, it is provided in the alternator through the lower distribution hose. It is then necessary to measure the purity of gas to the top of the alternator: the lower hose leads to the valve of command in the post office of distribution that is opened opportunely and closed.

In the same way, the puff of the meter of the purity is starting up in the alternator with the spray in start. When the purity of carbon dioxide gotten on the meter of purity is besides 75%, the feeding is stopped and hydrogen is by following introduces to its room.

By reason of its relatively weak weight in relation to the CO_2, gas hydrogen is provided with the help of the superior hose of distribution of the alternator.

The purity of gas hydrogen must be measured to the bottom of the alternator that is for it the valve of command is opened appropriately and closed. A regulator of pressure is installed between the hose of gas feeding and the station of gas hydrogen in order to maintain the pressure of the internal gas to a value wanted of 1 to 8 Abs bars.

In this application, we present on the one hand, the programming of the general numbering, timetable and daily of the gas hydrogen consumption and on the other hand, the configuration of a new tabular circuit gas hydrogen containing the new information [21].

Figure 12 presents the new tabular of the Hydrogen circuit containing the new modifications.

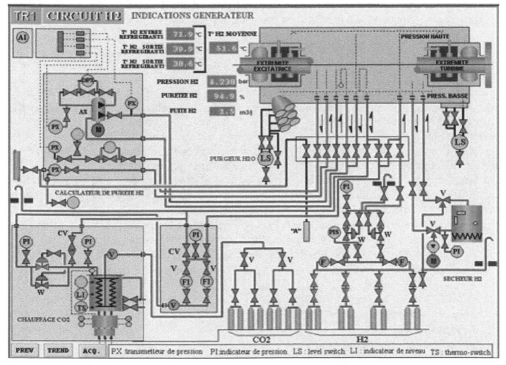

Fig. 12. New tabular of the hydrogen circuit [21].

The TPP of Radès arranges a regulator of pressure that assures the feeding of H2. When the uncommunicative gas in the alternator and the pressure is increased the regulator in will be stopped. The calculation of hydrogen flights makes himself by hand therefore we cannot have an exact value on these flights. This value is not displayed on the SCADA system.

The daily taking of values of numberings of flights of hydrogen consumption decreases the precision of calculation of the output of every slice.

To remedy these problems we propose a solution of automatic reading of the value of the gas hydrogen flight. The proposed solution is to make the calculation of flights by the SCADA system and to program blocks of the daily calculation. This solution is automatic and cyclic where the period of the time is stationary.

The algorithmic of treatment is based on the concepts of block and diagram (or compound). Indeed, a block is a software entity that achieves a specific function more less complex (stake to the ladder, conversion, filtering, calculation, test of alarms, etc.) definite by its algorithm.

For the configuration stage, we used the ICC software (Integrated Control Configuration). This software permits the creation and configuration of the resident program in the CP60.

For the conception stage of the tabular, we used the FoxDraw software. This software possesses a library of components permitting to represent the various elements of an industrial installation.

5. Application of the supervision of a water steam - cycle a TPP

Considering that the water used contains an elevated rate in dissolved salts and in matter suspended, it is indispensable to adopt a stage of pretreatment to assure the good working of the inverse osmosis installation and to protect modules against risks of usuries, corrosion and especially membrane calmative.

The pretreatment is constituted of two filtration chains each including a sand filter and an active coal filter. Thereafter, we present the two stations of the TPP: inverse osmosis and demineralization.

The control of the water quality is an important task to maintain the efficiency and the sure and continuous working of the power station. To guarantee the best water quality at the level of the water steam circuit, the TPP of Radès arranges an inverse osmosis station that permits to eliminate the majority of salts dissolved in the raw water before being treated in a demineralization station (See Figure 13). This stage serves to minimize risks of failing by corrosion of the turbine or the loss of the efficiency and the power.

The basic principle of the ion exchange consists in withdrawing ions (remaining salts that are lower to 8%) in solution in water is to recover an ion of value, either to eliminate a harmful or bothersome ion for the ulterior utilization of water.

The exchange of ions is a process which ions with a certain load contents in a solution are eliminated of this solution, and replaced in the same way by an equivalent quantity of other ions load gave out by the strong but the opposite load ions are not affected.

In the demineralization chain, osmosis water passes by the following stages:

- a weak cationic exchanger (CF1);
- a strong cationic exchanger (CF2);
- a weak anionic exchanger (AF1);
- a degasser;
- a strong anionic exchanger (AF2);
- a strong cationic exchanger (CF3);
- a strong anionic exchanger (AF3).

After the demineralization, the water must have a lower conductivity of 0.2 μS/cm, a pH between 6.5 and 7.5; silica < 30 ppb.

Figure 14 presents the cycle of the water treatment in the demineralization station.

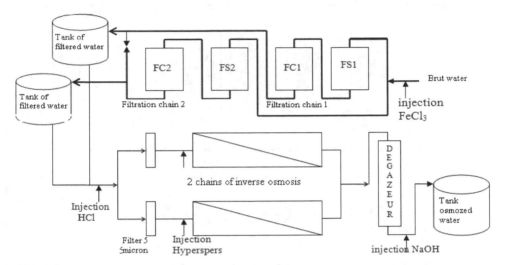

Legend:
FS1: Sand filter of the filtration chain 1.
FC1: Active coals filter of the filtration chain 1.
FS2: Sand filter of the filtration chain 2.
FC2: Active coals filter of the filtration chain 2.
The bold lines present the water circuit in the two filtration chains and the light lines present the water circuit in the two inverse osmosis chains.

Fig. 13. Functional diagram of the inverse osmosis station of the TPP.

In order to remedy to the absence of indication, of follow-up and of storage of the chemical characteristics of the water of the furnace, it is necessary to achieve an interfacing between the chemical sensors (pH meter and conductivity meter) and the stations of surveillance of the control room of the TPP.

The interfacing of the signals of the pH and the conductivity of the ball furnace is assured by a data configuration of both analogical and numeric signals and requires a unique code for every entrance which must be programmed in the data base system.

This application is declined in six stages:

Stage 1: Choosing the site of the signal (FBM module).

Stage 2: Programming both AIN and CIN blocks for the supervision of the signals pH (4 to 20 mA) and conductivity (alarm).

Stage 3: Testing both AIN and CIN blocks by injection of current and by short circuit.

Stage 4: Passing the cable between the sampling room and the SCADA room.

Stage 5: Connecting the signals in the two modules 10FBM215 and 10FBM325.

Stage 6: Conceiving a new tabular for the general vision of the sampling room.

The last stage of this application of interfacing consists in improving the tabular pH meter and conductivity meter.

Figure 15 presents the display of the sampling room containing the chemical analysis parameters of the water - steam cycle of the TPP.

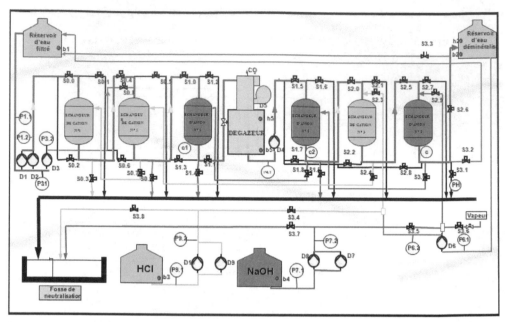

Fig. 14. Demineralization station of the TPP.

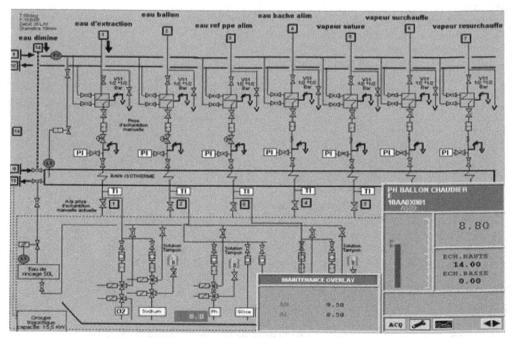

Fig. 15. Display of the chemical analysis parameters [22].

6. Application of the supervision of pumps vibrations of a TPP

Systems of vibration surveillance are often equipped of measure chains for other complementary parameters, as the axial position, the crankiness, the differential dilation, the dynamic pressure, the speed of rotation and the temperature.

Among the new systems of measures, we mention notably IDS (system of icing detection) and AGMS (system of measure of the bore between the rotor and the stator) that complete a system of vibration surveillance efficiently, but that are also usable as of the autonomous specific systems [23].

The MMS system (System of Machine Surveillance) is the synthesis of the long experience of Vibro-Meter in the domain of the surveillance of machines and its expertise to master technologies of vanguard as for the manufacture of the electronic of surveillance.

The instrument of vibration control measures the vibration all the time when machines (turbine of power plant, big dimension compressor, pump, blower...) are in service. When the supervised vibration reached the amplitude of vibration, that is adjusted in advance, the instrument gives out an exit of point of alarm contact to give a warning to the working of the machine or gives out an instruction to stop the working of the machine, avoiding so the danger and accidents before they occur.

The mechanical vibration that is developed in a machine is controlled by a sensor of vibration and is converted in electric signal and this signal is introduced in an amplifier of vibration. In this amplifier, a signal that is proportional to the speed of vibration and supervised by an instrument of vibration control, and convert in a signal that is proportional to the displacement of the vibration, and this last is to its tower convert in a tension to continuous current, that is given back like signal to an indicator and a signal to the circuit of alarm.

The instrument of vibration measure used in our application is constituted by a sensor of vibration (Model U1-FH) and an instrument of vibration control (Model AVR-148). In fact, the sensor of vibration, model U1-F, is similar to the construction of a loudspeaker to permanent magnet. The sensor is attached to the machine on the one hand with screws and on the other hand to connect to the system of registration with the special cables.

With sensors of Vibro-Meter, we can measure in general most the critical parameters in the surveillance of machines, but particularly what concerns vibrations. In this domain, Vibro-Meter proposes a vast range of sensors, of conditioners of the signal as well as an effective signal transmission.

Sensors of proximity and other translators of displacement are based on currents of Foucault and present a high linearity with an active compensation of the temperature. Some sensors, as piezoelectric accelerometers, have favors to be deprived of the mobile pieces, what permits to guarantee their reliability and a long life span.

The piezoelectric sensor is used as detector of shock, vibration or percussion. It captures the mechanical vibrations that transmit itself in a material.

Figure 16 presents the diagram block of the system vibratory surveillance of a pump used in a TPP [24].

To achieve a complete monitor of surveillance, we always associate a module of treatment UVC 691 with a module of surveillance with a high performance PLD 772.

UVC 691 is a module of signal treatment assorted to different sensors and conditioners via a galvanic separation.

Most modules of Vibro-Meter provide unipolar signals in the range of 0 to 10 V DC. However, the PLD 772 can accept some bipolar signals in the range of 0 to ±10 V DC.

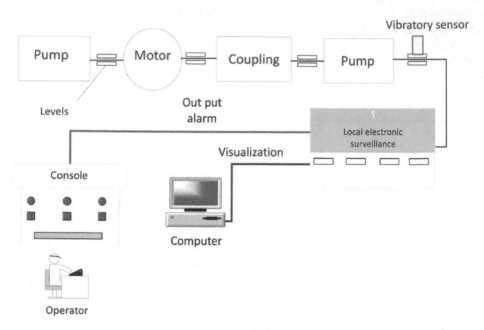

Fig. 16. Block diagram of vibratory surveillance.

In fashion of programming of the PLD 772, the user has the possibility to define the calibration of the display and all parameters of alarm. While equipping the PLD 772 of an interfacing RS-485, the module is capable to the digital communication (Figure 17). Thus, a system of surveillance can make part of a cabled network. A computer detains the main computer role. All other modules PLDS 772 in racks are some secondary stations. Such a link between a system of surveillance and a main computer is in measure to do functions of programming from afar and of data transfer.

The central computer can read the calibration of every module at all times PLD 772. Instructions of set up and a special authorization permit to modify parameters of calibration or doorsteps of alarm of every surveillance module. Commands become thus easy and the result is from afar a fully programmable surveillance system.

The computer calls each module periodically to ask it the measured values (DC signals) and the state of alarm of every channel. Such a process of acquirement suits the registration of data and the creation of a data basis very well with the acquisition in DC in order to do an analysis of tendency subsequently using software of conditional maintenance.

We have built many displays with FoxDraw that are used by FoxView, and become the I/A Series interface to the pumping process of the TPP (See Figure 18).

Monitoring display and diagnosis display are important for the supervisory system. A global view of the system should appear the information needed for the pumping process [25].

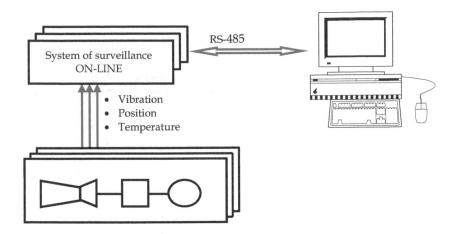

Fig. 17. Transfer of data between units.

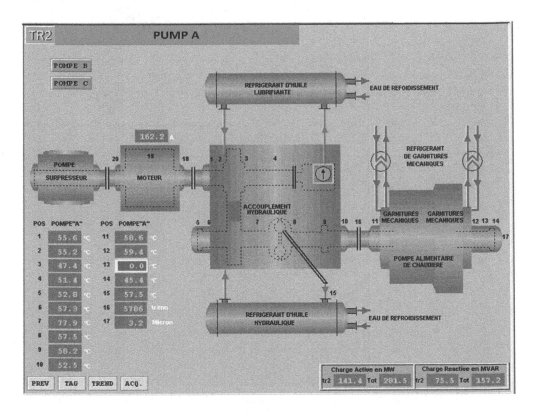

Fig. 18. Display of the pump A of the TPP [24].

7. Conclusion

SCADA systems are used to control and monitor physical processes, examples of which are transmission of electricity, transportation of gas and oil in pipelines, water distribution, traffic lights, and other systems used as the basis of modern society.

SCADA system gathers information, transfers the information back to the central site, alerting the home station and carrying out necessary analysis and control, and displaying the information in a logical and organized fashion.

In this work, we presented on the one hand an application of Functional Analysis (FA) techniques on a SCADA system of thermal power plant (TPP) and on the other hand some examples of SACDA applications.

The first application of the SCADA system consists in integrating a module for the calculation of hydrogen flights in the alternator. The proposed solution was permit to elaborate a new tabular for the hydrogen circuit containing the new modifications. The second application of the SCADA system consists in interfacing the chemical analysis parameters of a water- steam cycle to the SCADA system. The third application is related to the supervision of a system of vibratory surveillance in a TPP. This application enables us the creating and the maintaining dynamically updating the pumping process displays.

This achieved applications is going to facilitate the work of both laboratory and instrumentation agents in the TPP.

Functional Analysis techniques seem to be a promising way because the major advantage of these kinds of techniques is due to the concept of function and abstraction hierarchy which are familiar to the human operator. These techniques permit the complexity of a system to be overcome.

In this paper, the application of SADT and SA-RT methods on a real system, a SCADA system of a TPP generates a source of useful information for the design of a supervisory system (monitoring and diagnosis displays, definition of alarms, etc.). So, research into the application of FA techniques for the design of a human centered supervisory system must be intensified in order to solve several difficulties and to improve their efficiency (tools to build the model, tools to check the validity of the model, etc.).

8. Acknowledgements

The author extends his especial thanks to Moncef Harrabi and Moez Lakhoua, Tunisian Society of Electricity and Gas (STEG) for their valuable help.

9. References

[1] Bailey D., Wright E., *Practical SCADA for Industry*, Elsevier, 2003.
[2] Clarke G., Reynders D., Wright E., *Practical Modern SCADA Protocols*, Elsevier, 2003.
[3] Gergely E.I., Coroiu L., Popentiu-Vladicescu F., Analysis of the Influence of the Programming Approach on the Response Time in PLC Control Programs,

Journal of Computer Science and Control Systems, 3(1), 2010, Oradea, Romania, pp 61-64.

[4] Avlonitis S.A., Pappas M., Moutesidis K., Avlonitis D., Kouroumbas K., Vlachakis N., *PC based SCADA system and additional safety measures for small desalination plants*, Desalination 165, 2004, p. 165-176.

[5] Heng G. T., *Microcomputer-based remote terminal unit for a SCADA system*, Microprocessors and Microsystems, Volume 20, Issue 1, 1996, p. 39-45.

[6] Horng J.H., *SCADA system of DC motor with implementation of fuzzy logic controller on neural network*, Advances in Engineering Software 33, 2002, p. 361–364.

[7] Patel M., Cole G. R., Pryor T. L., Wilmota N. A., *Development of a novel SCADA system for laboratory testing*, ISA Transactions 43, 2004, p. 477–490.

[8] Munro K., *SCADA - A critical situation, Network Security*, Vol. 2008, Issue 1, 2008, p. 4-6.

[9] Wiles J., *Supervisory Control and Data Acquisition: Techno Security's Guide to Securing SCADA*, 2008, p. 61-68.

[10] Wiles J., *Techno Security's Guide to Securing SCADA: A Comprehensive Handbook On Protecting The Critical Infrastructure*, Elsevier, 2008.

[11] STEG, *Rapport Annuel de la STEG*, Société Tunisienne de l'Electricité et du Gaz, 2008.

[12] Revue de l'électricité et du gaz, Société Tunisienne de l'Electricité et du Gaz, N°16, 2008.

[13] Peffers K., Planning for IS applications: a practical, information theoretical method and case study in mobile financial services, Information & Management, Vol.42, Issue 3, March 2005, pp. 483-501.

[14] Marca DA., McGowan CL., SADT: structured analysis and design technique, New York: McGraw-Hill Book Co., Inc., 1988.

[15] Schoman K., Ross D.T., Structured analysis for requirements definition, *IEEE Transaction on Software Engineering* 3 (1), pp. 6-15, 1977.

[16] Lakhoua M.N, *Application of Functional Analysis on a SCADA system of a Thermal Power Plant*, Advances in Electrical and Computer Engineering journal, Issue No2. Vol.9, 2009, p. 90-98.

[17] Lakhoua M.N, *Application of functional analysis for the design of supervisory systems: Case study of heavy fuel-oil tanks*, International Transactions on Systems Science and Applications, Vol.5, N°1, 2009, p.21-33.

[18] Gomaa H., Software design methods for the design of large-scale real-time systems, Journal of Systems and Software, vol. 25, Issue 2, pp 127-146, 1994.

[19] Jaulent P., Génie logiciel les méthodes : SADT, SA, E-A, SA-RT, SYS-P-O, OOD, HOOD, Armand Colin, Paris, 1992.

[20] Lakhoua M.N., Application of SA-RT method to supervisory systems, Journal of Electrical Engineering, ISSN: 1582-4594, Politechnica Publishing House, vol.11, N°2, 2011.

[21] Lakhoua M.N., Application of a SCADA system on a hydrogen station, Journal of Electrical Engineering, vol.10, N°3, 2010.

[22] Lakhoua M.N., "Systemic analysis of a supervisory control and data acquisition system", Journal of Electrical Engineering, ISSN: 1582-4594, Politechnica Publishing House, vol.11, N°1, 2011.

[23] Système de surveillance des machines MMS, VIBRO-METER SA, Fribourg/Suisse, 4, 1990.

[24] Lakhoua M.N, *Surveillance of pumps vibrations using a SCADA*, Control Engineering and Applied Informatics, Vol.12. N°1, 2010.

[25] Lakhoua M.N., Application of a supervisory control and data acquisition system in a thermal power plant, 12th symposium Large Scale Systems: Theory and Applications LSS'10, 11-14 July 2010, France.

7

Adaptive Gas Path Modeling in Gas Turbine Health Monitoring

E. A. Ogbonnaya[1], K. T. Johnson[1], H. U. Ugwu[2],
C. A. N. Johnson[3] and Barugu Peter Forsman[4]

[1]*Department of Marine Engineering, Rivers State University of Science and Technology,*
Port Harcourt
[2]*Department of Mechanical Engineering, Michael Okpara University of Agriculture,*
Umudike-Umuahia
[3]*Department of Marine Engineering, Niger Delta University, Wilberforce Island,*
Bayelsa State
[4]*Department of Welding, Oil and Gas Engineering, Petroleum Training Institute,*
Effurun, Delta State
Nigeria

1. Introduction

The ability to model the behavior of gas turbines (GTs) is critical in all aspects of energy and power generation engineering. A computerized approach giving the possibility for a more detailed gas path component fault diagnosis and prognosis using the MVR is presented. A diagnostic engine performance model is the main tool that points to the faulty engine component. The diagnostic component model was also used to come up with the software code-named Thermodynamics and Performance Condition Monitoring(THAPCOM) written in C++ programming language to effectively identify the fault on the engine. Several scheduled visits were thus made to AFAM IV, GT 18, TYPE 13D power plant located near Port Harcourt, in Rivers State of Nigeria. Continuous and periodic monitoring of the thermodynamics/performance parameters such as temperature, pressure, air pumping capability, rotational speed, air, fuel and gas flow were carried out. This exercise lasted for a period of three months on hourly basis to predict the health of the engine. When these data were analyzed by the software, the following results were obtained $\frac{\Delta A_N}{A_N} = 1.4598e^{-0.008}$, $\frac{\Delta n_c}{n_c} = 1.6630e^{-0.007}$, $\frac{\Delta T_c}{T_c} = 1.1626e^{-0.008}$ and $\frac{\Delta T_{3c}}{T_{3c}} = 7.5508e^{-0.007}$, which correspond to average overall efficiency of 27.3% and active load of 48MW. These were indications that the test engine had suffered from fouling, degraded compressor performance and seal leakage. THAPCOM gives an alarm signal when a set limit is exceeded so that maintenance could be scheduled.

Nomenclature and Abbreviation:

A = Actual value
A_L = Active load (MW)
A_N = Area of nozzle (m²)

Δ = Difference between actual and reference value
η_T = Isentropic efficiency of turbine (IET)
MVR= Multi-variable response
n_c = Isentropic efficiency of compressor (ICE)
N = Shaft speed (RPM)
η_o = Overall cycle efficiency
D = Percentage deviation a_1, a_2...a_{16} are values of the coefficients
P_1 = Compressor inlet pressure (bar)
P_2 = Compressor exit pressure (bar)
R = Reference value
T_1 = Compressor inlet temperature (k)
T_2 = Compressor exit temperature (k)
T'_2 = Isentrropic compressor exit temperature (k)
T_3 = Turbine inlet temperature (k)
T_4 = Turbine exhaust temperature (k)
T'_4 = Turbine is isentropic exhaust temperature (k)
Γ_c = Air pumping capability of compressor, APC (m³/s)
W_F = Fuel flow (kgls)
γ = $\dfrac{Specific\ heat\ at\ constant\ pressure}{Specific\ heat\ at\ constant\ volume}$ = 1.4

In today's competitive business environment and low profit margins, manufacturers are faced with the growing production demands while cutting the cost of manufacturing (Bell, 2003). One pervasive cost that drags down productivity is the unplanned equipment and manufacturing down time. High performance turbo machines especially GTs are now extremely important elements of industries. Some areas where GTs are used include electric power, petrochemical, mining, marine, air craft, onshore and offshore oil and gas industries. (Ogbonnaya (2004a), Ogbonnaya (2004b), Pussey (2007), Brun and Kurz (2007), Rieger, et al, (1990), Ogbonnaya (2009), Ogbonnaya and Theophilus–Johnson (2010), Loboda and Yepifanov (2010), Aretakis, et al, (2010)). This chapter is therefore timely for these maritime organisations to adopt the proactive measures being preferred to prevent their equipment from catastrophic downtime. While running, GTs are adversely affected by the environmental factors such as dust particles, smoke, smog, oil mist and high humidity (Brun and Kurz (2007), Schneider, et al, (2009), Ogbonnaya and Theophilus-Johnson (2010)). These factors have resulted to degradation mechanisms of fouling, erosion, corrosion and abrasion which reduce the overall performance of the aerodynamic components of the plant (Brun and Kurz (2007), Schneider, et al, (2009), Ogbonnaya and Theophilus-Johnson (2010)).

The use of performance engine models in diagnostics has been initiated since the early 70's. The first approaches were based on linearized models (Aretakis, et al, 2010) while Stamatis, et al., (1991), introduced the concept of using directly non-linear models in diagnostics. Also, methodological steps of simulation and modeling used by Maria (1997), Erbes, et al., (1993), Erbes and Palmer (1994) and Ogbonnaya (2004a) proved that modeling and simulation are handy tools for condition monitoring.

The gas path analysis technique gives the possibility to identify the amount of deterioration of individual components and assess its effect on overall performance providing information, which is valuable for improving cost effectiveness of maintenance actions. An analytical tool that can be used for this purpose was presented in Doel (1994). Performance diagnostic methods for identifying deterioration has also been presented by Urban and Volponi (1992) and Volponi (1994). These approaches show which component is malfunctioning and depending on the established experience, can offer an evaluation of the nature of malfunction. The compressor and turbine deterioration are the main cause of the overall performance deterioration. The introduction of measured gas path variables such as pressure, temperature, rotational speed, fuel flow, air flow, gas flow etc is hereby consolidated through this project.

The gas path components, such as compressors, turbines and combustion chamber can be affected by foreign object damage, fouling, tip rubs, seal wear and erosion. The ability to identify the faulty component and simultaneously diagnose the defect with its consequences is another purpose of this chapter. It also allows the operator to take necessary maintenance measures to rectify the fault and provide an assessment of the GTs life cycle and valuable data for prognostics and condition based maintenance scheduling. To achieve these, a detailed component diagnostic modeling needs to be applied. Therefore, the technology of prognosis is recommended in this work because it involves diagnosis, condition and failure model. Prevention of catastrophic and unexpected downtime was thoroughly considered to come up with the software called "THAPCOM" written in C++ programming language to diagnose and prognose the health of the GT. Trend monitoring technique was applied using multiple variable mathematical models (MCMV) in matrix form (Bently et al., 2002). The introduction of "THAPCOM" into GT diagnosis and prognosis conforms to the use of thermodynamics / performance parameters (dependent and independent parameters) as it is the driving force of the GT. "THAPCOM" stands for THermodynamics And Performance COndition Monitoring. As stated in Uhumnwangho, et al., (2003), Brun and Kurz (2007) and Ogbonnaya (2009), the deviation of GT thermodynamics and performance parametric values from their reference values stated in the manufacturer's manual is an indication of impending failure. This is because condition monitoring is the process of ascertaining the state of a parameter in an equipment such that any adverse significant deviation/change is an indication of impending failure.

1.1 Approaches to monitoring and data collection

Recently, continuous and periodic monitoring are used for GT data collections. Although, the presence of continuous monitoring does not eliminate the need for periodic monitoring (Guy, 1995), the continuous monitoring system warns the operator about imminent problems. Periodic monitoring along with the collection of external data provides a means for analysis and projection of potential long-term problem with respect to maintenance and operation (Ogbonnaya and Theophilus-Johnson, 2010). The collection of GT model data is capable of acquiring the necessary information to monitor and trend the engine health. This present work also utilized periodic monitoring to achieve its aim.

1.2 Brief condition monitoring methods

The already known novel methods of gas path model-based component condition monitoring used by Ogbonnaya (1998, 2004a) is integrated into the THAPCOM. The works of Loboda and Yepifanov, (2010); Donald, et al, (2008); Loboda, (2008); Fast, et al., (2009); Aretakis, et al; (2003); Romesis and Mathioudakis, (2003); Roemer and Kacprzynski, (2000); Volponi, et al., (2003); Kamboukous and Mathioudakis, (2005) on gas path analysis in condition monitoring, were also critically considered to actualize this task. More so, trend monitoring as utilized in Bently, et al., (2002) and Uhumnwangho, et al., (2003) was also seen as a viable tool for this package. Finally, the benefits of motor condition monitoring (MCM) in Bell (2003); Pussey (2007) were rigorously brought to bear to bring this present research to fruition. In most of these works, as well as Ogbonnaya and Koumako (2006), operational safety and control of the GT engines were equally harnessed in this new technology.

1.3 The software "THAPCOM"

THAPCOM is a viable diagnostic tool because it is capable of providing early warning to progressively indicate imminent fault during engine operation. It analyses conditions to prevent unplanned down time. THAPCOM is an inexpensive diagnostic tool that gives accurate maintenance decision information which is understandable to both low and semi-skilled personnel. Therefore, it also eliminates the short-comings of both performance and trend monitoring. Their similarity is that they all measure pressure, flow temperature and rotational speed simultaneously. The plus of THAPCOM is that it relates deterioration to consequences. THAPCOM uses model-based fault detection and diagnostic techniques (Ogbonnaya and Theophilus-Johnson (2011). This relates the deterioration which the engine has undergone to consequences along the gas path of a GT engine. When THAPCOM is interfaced with a GT, it first studies the system for a period of time through acquiring and processing the real-time data from the engine. The data is processed using system identification algorithms for both the actual (operational) behavior to the reference (design) behavior of the engine.

THAPCOM stores the processed data in its internal data base and also serves as the reference (design) values. These reference values are usually mean values of the performance parameters during factory test. During the monitoring session, THAPCOM processes the acquired engine data and compares the results with the data stored in its internal database. If the results obtained from the acquired data are significantly different from the reference values, THAPCOM indicates a faulty level through a series of alarm signal. The level is determined by the magnitude of their percentage deviation when compared. THAPCOM monitors, compares 15 thermodynamics and performance parameters and uses 4 of the parameters to obtain the coefficients. THAPCOM is similar to MCM, ANNs used in Ogbonnaya (2004a and 2009) in their mode of operation but their difference is that MCM measures only current and voltage while THAPCOM measures thermodynamics and performance parameters. ANNs was used to diagnose and prognose GT rotor shaft faults. THAPCOM displays the most sensitive performance parameters of the engine such as those which are used for diagnostics and prognostics. It is an advancement of the component model-based condition monitoring for a GT engine (Ogbonnaya et al, 2010).

2. Multi variable mathematical modeling

The approach used in this research is trend monitoring as MVR in matrix form. Data were obtained both statistically and analytically and constitute the most sensitive thermodynamics and performance parameters at the various components of the engine. Data were collected on hourly basis, for a period of three months from an operational GT plant used for electric power generation. The data were sampled and the mean taken for weekly basis. The GT is a 75 MW plant. This technique is in accordance with the methods stated in subsection (1.2). For instance, the method of model-based computer program yielded accurate results than the manual method. The method of model-based computer programming is faster in diagnosing faults. This use of computer program approach, signals the limit of operation through instrumentation in the form of alarm (Baker, 1991; Bergman, et al, 1993; Stamatis et al, 2001; Alexious and Mathioudakis, 2006; Ogbonnaya et al, 2010). This present work would contribute solution to the unexpected failure/down time of GTs by giving timely alarm signals. The deviations of the thermodynamics and performance parameters when the actual values were compared to their reference values will be used to analyze the MVMMs to diagnose and prognosis the health of the GT. The data collected from the test engine was obtained using the following model thermodynamics equations. It was assumed that $P_1 = P_4$ and $P_2 = P_3$.

Isentropic compression of the compressor was obtained as follows:

$$\frac{T'_2}{T_1} = \left(\frac{P_2}{P_1}\right)^{\frac{\gamma-1}{\gamma}} \tag{1}$$

Similarly, isentropic expansion of the turbine was obtained as follows:

$$\frac{T_3}{T'_4} = \left(\frac{P_3}{P_4}\right)^{\frac{\gamma-1}{\gamma}} \tag{2}$$

Isentropic efficiency of compressor = $\dfrac{Isentropic\ Enthalpy\ Drop}{Actual\ Enthalpy\ Drop}$

$$\eta_c = \frac{T_{2^1} - T_1}{T_2 - T_1} \tag{3}$$

Isentropic efficiency of turbine $\quad = \quad \dfrac{Actual Enthalpy\ Drop}{Isentropic\ Enthalpy\ Drop}$

$$\eta_\tau = \frac{T_3 - T_4}{T_2 - T'_4} \tag{4}$$

While the following model deviation equations were applied

$$\frac{\Delta T_3}{T_3} = \frac{T_{3A} - T_{3R}}{T_{3R}} \tag{5}$$

$$\frac{\Delta N}{N} = \frac{N_A - N_R}{N_R} \tag{6}$$

$$\frac{\Delta \eta_c}{\eta_c} = \frac{\eta_{cA} - \eta_{cR}}{\eta_{cR}} \tag{7}$$

$$\frac{\Delta \Gamma_c}{\Gamma_c} = \frac{\Gamma_{CA} - \Gamma_{CR}}{\Gamma_{CR}} \tag{8}$$

The parameters in Equations (5) to (8) are the independent variables in the MVMMs. These equations were used to generate the coefficients a_1 to a_{16} in the MVMMs. a_1 to a_{16} are expressed as functions of:

$$\left.\begin{array}{rcl} \dfrac{\Delta T_3}{T_3} &=& f[a_1,\ a_5,\ a_9,\ a_{13}] \\[2ex] \dfrac{\Delta N}{N} &=& g[a_2,\ a_6,\ a_{10},\ a_{14}] \\[2ex] \dfrac{\Delta \eta_c}{\eta_c} &=& h[a_3,\ a_7,\ a_{11},\ a_{15}] \\[2ex] \dfrac{\Delta \Gamma_c}{\Gamma_c} &=& i[a_4,\ a_8,\ a_{12},\ a_{16}] \end{array}\right\} \tag{9}$$

$$f, g, h, i = F_n \text{ Parameters} \tag{10}$$

The significance of this approach is based on the interface between the components of air and gas path. This approach considered the analysis in terms of the measurable dependent data and the independent performance parameters calculated by a mathematical model based on engine thermodynamics.

3. Dependent and independent variables

The independent and dependent parameters represent the variables in various engine components thermodynamics relationship such as the compressor, combustor and turbine units (Bently, et al., 2002). The differential and manipulation of these equations allow the derivation of a general relationship between each change in a dependent parameter and its resulting effects on each independent parameter in turn with all other variables held constant. A matrix was formed using these coefficient relationships by superposition of the independent variable on each independent parameter. The independent parameters are T_3, N, η_C, and Γ_C, while the dependent parameters are P_2, T_2 W_F and A_n. A combination of the MVMMs constitute a 4 x 4 matrix in which the variables are related by the constant coefficients a_1 to a_{16}. This matrix was evaluated as a 4 x 3 matrix holding the speed constant in turn to generate each independent parameter change (Bentley, et al., 2002). This is shown in equation (11).

4. The flowchart for the simulation

By substituting equation (12) into (14), $\frac{\Delta A_N}{A_N}$ can be obtained. Equations (5) to (8) and (14) were used for the simulation of THAPCOM in C++ programming language to proactively monitor the health of the GT. The flowchart drawn from these equations is presented in figure 1. It is from this flowchart that a computer program in C++ used to actualise the work is written. The most salient feature of THAPCOM flowchart and program is that it has two subroutines for ease of manipulation.

The input subroutine in the flowchart helped to store values of T_1, T_2, P_1, P_2, T_3, N, T_4, Γ_C, W_F and L. These values were later returned in subsequent parts of the program where they were needed and used to compute $\frac{\Delta T_3}{T_3}$, $\frac{\Delta \eta_C}{\eta_C}$ and $\frac{\Delta \Gamma_C}{\Gamma_C}$. This was done after individual values of η_T, η_C, $\eta_0 \ldots$ were computed.

$$\begin{array}{c}
\begin{array}{cccc}
\frac{\Delta T_3}{T_3} & \frac{\Delta N}{N} & \frac{\Delta \eta_C}{\eta_C} & \frac{\Delta \Gamma_C}{\Gamma_C}
\end{array} \\
\begin{array}{c}
\frac{\Delta P_2}{P_2} \\[2mm]
\frac{\Delta T_2}{T_2} \\[2mm]
\frac{\Delta W_F}{W_F} \\[2mm]
\frac{\Delta A_N}{A_N}
\end{array}
\left.\begin{array}{cccc}
a_1, & a_2, & a_3, & a_4 \\[2mm]
a_5, & a_6, & a_7, & a_8 \\[2mm]
a_9, & a_{10}, & a_{11}, & a_{12} \\[2mm]
a_{13}, & a_{14}, & a_{15}, & a_{16}
\end{array}\right\}
\end{array} \tag{11}$$

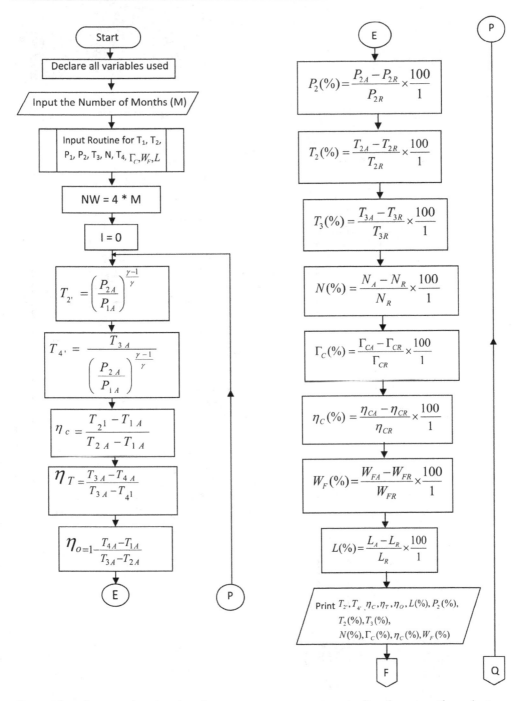

Fig. 1. Flowchart used to develop the computer program to actualize the gas path analysis

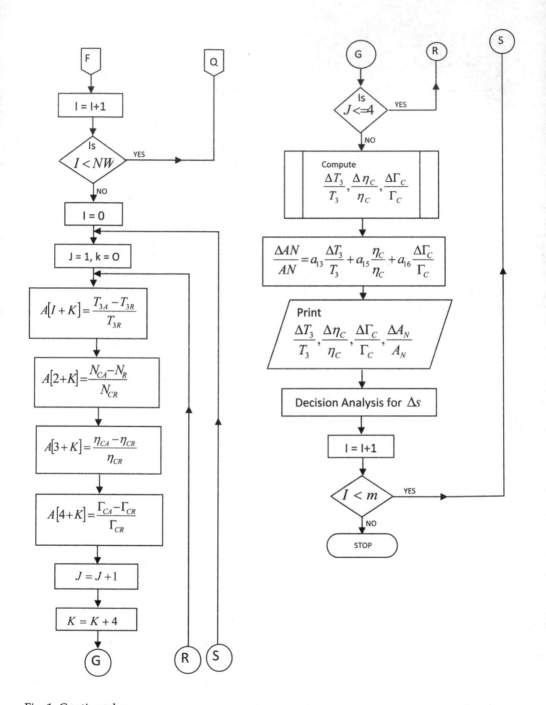

Fig. 1. Continued

The independent parameters are evaluated as follows:

$$
\frac{\Delta T_3}{T_3} = \begin{vmatrix} a_2, & a_3, & a_4 \\ a_6, & a_7, & a_8 \\ a_{10}, & a_{11}, & a_{12} \\ a_{14}, & a_{15}, & a_{16} \end{vmatrix}
$$

$$
\frac{\Delta \eta_C}{\eta_C} = \begin{vmatrix} a_1, & a_2, & a_4 \\ a_5, & a_6, & a_8 \\ a_9, & a_{16}, & a_{12} \\ a_{13}, & a_{14}, & a_{16} \end{vmatrix} \qquad (12)
$$

$$
\frac{\Delta \Gamma_C}{\Gamma_C} = \begin{vmatrix} a_1, & a_2, & a_3 \\ a_5, & a_6, & a_7 \\ a_9, & a_{10}, & a_{11} \\ a_{13}, & a_{14}, & a_{15} \end{vmatrix}
$$

Equations (11) and (12) show how the dependent variables were obtained:

$$
\frac{\Delta P_2}{P_2} = a_1 \frac{\Delta T_3}{T_3} + a_3 \frac{\Delta \eta_C}{\eta_C} + a_4 \frac{\Delta \Gamma_C}{\Gamma_C}
$$

$$
\frac{\Delta T_2}{T_2} = a_5 \frac{\Delta T_3}{T_3} + a_7 \frac{\Delta \eta_C}{\eta_C} + a_8 \frac{\Delta \Gamma_C}{\Gamma_C} \qquad (13)
$$

$$
\frac{\Delta W_F}{W_F} = a_9 \frac{\Delta T_3}{T_3} + a_{11} \frac{\Delta \eta_C}{\eta_C} + a_{12} \frac{\Delta \Gamma_C}{\Gamma_C}
$$

$$
\frac{\Delta A_N}{A_N} = a_{13} \frac{\Delta T_3}{T_3} + a_{15} \frac{\Delta \eta_C}{\eta_C} + a_{16} \frac{\Delta \Gamma_C}{\Gamma_C} \qquad (14)
$$

With a view to actualize MVMMs, the data shown in tables 1(a) and (b) were collected from the operational GT plant. Figures 2 and 3 are the graphs of percentage deviation in P_2 and T_2 against date in weeks while a combined graph of percentage deviation in P_2, T_2, Γ_C and T_3 are shown in figure 4.

5. Implementation

The coefficients of each performance parameters are depicted in equation (9) in relation to equations (5) to (8), when the actual value is compared to the reference value. When these coefficients are used with the MVMMs, to diagnose and prognose the GT faults, its state of health was made known. If, while trending its health using equations (12) and (14), and all the Δs = 0, with no performance change, then the GT is said to be healthy.

When $\frac{\Delta A_N}{A_N}$ = 0, $\frac{\Delta \eta_C}{\eta_C}$ and $\frac{\Delta \Gamma_C}{\Gamma_C}$ are downward and $\frac{\Delta T_3}{T_3}$ is upward, it implies degraded compressor. This is an indication of built up dirt, foreign object damage, blade erosion, missing blade, warped blade or seal leakage. The results of the simulation show that $\frac{\Delta A_N}{A_N}$ = $1.4598e^{-0.008}$, $\frac{\Delta \eta_C}{\eta_C}$ =$1.6630e^{-0.007}$, $\frac{\Delta \Gamma_C}{\Gamma_C}$ = $1.1626e^{-0.008}$ and $\frac{\Delta T_{3c}}{T_{3c}}$ = $7.5508e^{-0.007}$ for the first four weeks, since THAPCOM analyses data on cumulative basis. This showed that the GT had suffered from fouling, degraded compressor performance and seal leakage. Furthermore, figures 2 and 3 show the graphs of percentage deviation in compressor outlet pressure and temperature against date in weeks respectively. The table of values shows that the trajectories depict a sinusoidal trend. This is as a result of fouling, which is known for the reduction in compressor exit pressure from its design value. Figure 4 is a combined plot of P_2, T_2, T_3, N, Γ_C, η_C and A_L against date in weeks. It shows that, A_L suffered the highest deviation. Moreover figures 5, 6 and 7 show the path of percentage deviation in ICE, A_L and APC against date in weeks. The sinusoidal trend also means that compressor instabilities were setting in.

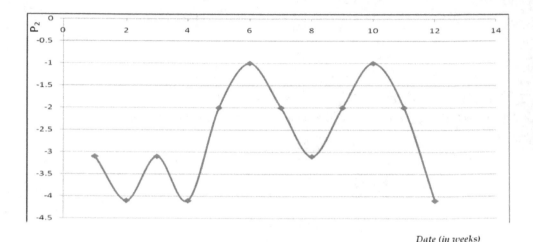

Date (in weeks)

Fig. 2. Percentage Deviation in P_2 against Date (in Weeks)

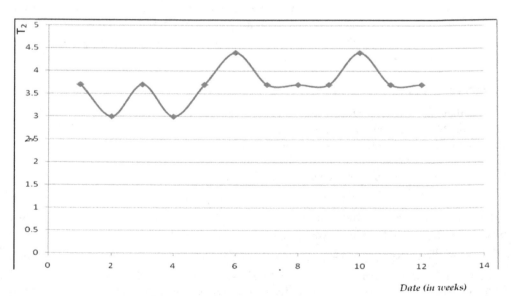

Date (in weeks)

Fig. 3. Percentage deviation in T_2 against date (in weeks)

Date Weeks	$T_1(k)$			$T_2(k)$			P_1 (bar)			P_2 (bar)			T_3 (k)			T_4 (k)			T_4^1 (K)	A_L (MW)		
	A	R	%D	A	R	%D	A	R	%D	A	R	%D	A	R (1)	%D	A	R	%D	A	A	R	%D
1.	299	300	-0.33	615.9	594	3.7	1.014	1.013	0.099	9.5	9.8	-3.1	1131	1223	-7.52	707	813	-13.04	596.8	48	75	-36.00
2.	298.1	300	-0.63	616.0	594	3.0	1.014	1.013	0.099	9.4	9.8	-4.1	1131	1223	-7.52	708	813	-12.92	598.6	49	75	-34.67
3.	299.1	300	-0.10	616.9	594	3.7	1.015	1.013	0.197	9.5	9.8	-3.1	1131	1223	-7.52	707	813	-13.04	597	49	75	-36.67
4.	299.7	300	-0.53	616.0	594	3.0	1.014	1.013	0.099	9.4	9.8	-4.1	1128	1223	-7.77	710	813	-12.67	597	48	75	-36.00

Date Weeks	W_F x10⁴ (kg/s)			T_2' (k)	η_C			η_T			η_O	$\Gamma_C (m^3/s)$			N (RPM)		
	A	R	%D	A	A	R	%D	A	R	%D	ηo	A	R	%D	A	R	%D
1.	4.96	6.0	-17.3	566.6	0.845	0.844	0.12	0.794	0.789	0.5	27.7	282	295	-4.4	3063	3000	2.1
2.	4.91	6.0	-18.2	563.2	0.853	0.844	0.12	0.795	0.789	0.6	27.8	290	295	-1.7	3061	3000	2.0
3.	4.70	6.0	-21.2	564.7	0.848	0.844	-0.59	0.794	0.789	0.5	27.8	280	295	-5.1	3056	3000	1.9
4.	4.67	6.0	-22.2	566.4	0.836	0.844	1.07	0.787	0.789	-0.2	27.0	295	295	0.0	3058	3000	1.9

Table 1. (a) Values of the thermodynamics and performance parameters taken from AFAM IV, GT18, TYPE 13D

Weeks	$\dfrac{\Delta T_3}{T_3}$		$\dfrac{\Delta N}{N}$		$\dfrac{\Delta \eta_c}{\eta_c}$		$\dfrac{\Delta \Gamma_c}{\Gamma_c}$	
1.	a_1	-0.075	a_2	0.021	a_3	0.0012	a_4	-0.044
2.	a_5	-0.075	a_6	0.020	a_7	0.0012	a_8	-0.017
3.	a_9	-0.075	a_{10}	0.019	a_{11}	-0.0059	a_{12}	-0.051
4.	a_{13}	-0.078	a_{14}	0.019	a_{15}	0.0107	a_{16}	0.000

Table 1. (b) Values of the ccoefficients

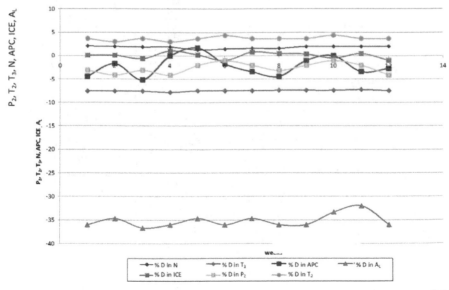

Date, Weeks

Fig. 4. Percentage deviation in P_2, T_2, T_3, N, APC, ICE and A_L against date (in weeks)

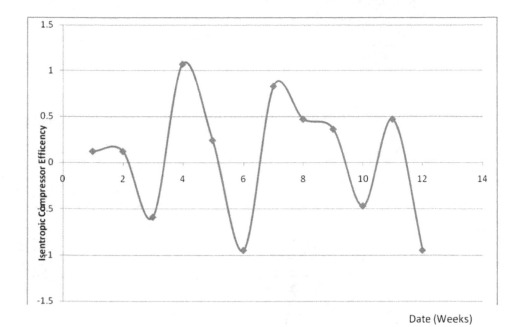

Fig. 5. Percentage deviation of ICE against date (in weeks)

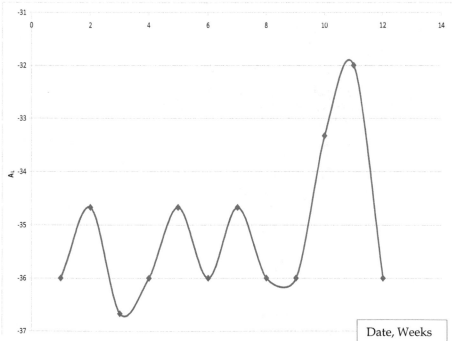

Fig. 6. Percentage deviation in A_L against date (in weeks)

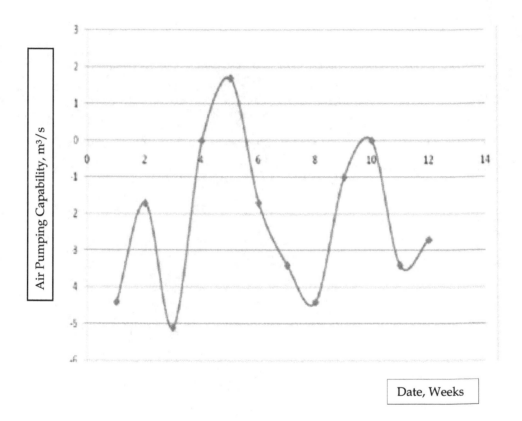

Fig. 7. Percentage deviation in APC against date (in weeks)

6. Conclusion

In this work, the MVMMs of a test engine was generated by taking advantage of the gas path analysis. The models were applied to develop the software "THAPCOM". This software thus enabled diagnosis and prognosis to be carried out on the equipment through the comparison between the actual and reference values of the engine. Advantage was brought to bear using previous works done on adaptive modelling of various aspects of GT health monitoring. The software when installed in a system interface of the GT enabled the proactive monitoring of the engine's health. The software gives an alarm signal whenever a set limit is near the dependent or independent parameters. This alarm signal allows the operator to carry out maintenance before the equipment fails.

7. Acknowledgements

The authors are highly grateful for the contributions of the staff of Afam Thermal Station where the data and experimentations were conducted. They are Engrs. D. U. Obiagazie, L. Ofurum, M. U. Ukpai, K. Kalio.

8. References

Alexious, A and Mathioudakis, K(2006) Gas Turbine Engine Performance Model Application Using an Objective Oriented Simulation Tool, *ASME Turbo-Expo 2006*, Power for Land, Sea and Air, The Barcelona, Spain, May 8-11. Available at *http://www.137.205.176.10/content/engine/sp²-asme-turbo-2006-alexious.pdf*

Aretakis, N., Roumeliotic, I and Mathioudakis, K(2010) Performance Model "Zooming" for In-Depth Component Fault Diagonsis, *Proceedings of ASME Turbo-Expo 2010, GT2010-23262*, Glasgow-Scotland, UK, pp. 1-2

Aretakis, N., Mathioudakis, K and Stamatis, A.(2003) Non- Linear Engine Component Fault Diagnosis From a Limited Number of Measurements Using a Combinational Approach, *Journal of Engineering for Gas Turbines and Power*, Vol. 125, Issue 3, pp. 642-650

Baker, W.E(1991) *Similarity Methods in Engineering Dynamics:Theory and Practice of Scale Modeling*(Revised Edition), ISBN :0-444-88156-5, Elsevier Science Publishers B.V, Amsterdam, The Netherlands, pp. 7-18

Bell, D.R., (2003): *The Hidden Cost of Downtime: Strategies for Improving Return on Assets*, Smart Signal Co. USA. pp. 1-4

Bergman,J.M., Boot, P and Woud, K. K(1993) Condition Monitoring of Diesel Engines with Component Models, *Paper 17 International Conference on Marine Environnmental and Safety (ICMES) 93, Marine Management(Holdings) Limited,*, The Netherlands.

Bently, D.E., Hatch, C.T., and Grisson, B., (2002): *Fundamentals of Rotating Machinery Diagnostics*, Bently Pressurized Bearing Press Co., Canada. 1st Print, p. 288

Brun, K., and Kurz, R., (2007): Gas Turbine Tutorial – Maintenance and Operating Practice Effects on Degradation and life, *Proceedings of the Thirty-sixth Turbo Machinery Symposium*, pp. 1-2.

www.igu.org/html/wgc2009/papers/docs/ *wgcFinal00076.pdf*

Doel, D(1994) A Gas Path Analysis Toolfor Commercial Jet Engines, *Transaction of ASME Journal. of Engineering for Gas Turbines and Power,* Vol.116, pp. 82-89

Donald, L.S., Volponi,A.J., Bird,J., Davison, C and Verson, R.E(2008) Benchmarking Gas Path Diagnostic Methods : Public Approach, *IGTI/ASME Turbo-Expo 2008, GT2008-51360,* Berlin- Germany, p.2.

Erbes, M. R., and Palmer, C. A., (1994): Simulation Mehtods used to Analyse the Performance of the GE E6541B Gas Turbine utilizing Low Heating Value Fuels. *ASME Cogen Turbo Power,* Portland Oregen. pp. 1 – 2.

Erbes, M. R., Palmer, C. A., and Pechti, P. A., (1993): Gas Cycle Performance Analysis of the LM2500 Gas Turbine utilizing Low Heating Values. IGTI – vol. 8. *ASME Cogen – Turbo Power.* pp .1-2.

Fast, M., Assadi, M., Pike, A and Breuhaus, P(2009) Different Conditon Monitoing Models for Gas Turbines by Means of Artifical Neural Networks, *IGTI/ASME Turbo-Expo2009, GT2009-59364,* Orlando- Florida, USA, p11

Guy, K. R., (1995): Turbine Generator Monitoring and Analysis, Mini-Course Notes, *Proceedings of Vibration Institute.* p2.
www.sandv.com /downloads/0703puse.pdf

Kamboukas, P and Mathioudakis, K(2005) Comparison of Linear and Non- linear of Gas Turbine Performance Diagnostics, *Journal of Engineering for Gas Turbines and Power,* Vol.127, Issue 1, pp...49-56.

Loboda, I(2008) Trustworthiness Problem of Gas Turbine Parametric Diagnosing, *5th IEAC Symposium of Technical and Safe Processes,* 2003, Washinton DC, USA, p.8.

Loboda, I and Yepifanov, S(2010) A mixed Data-Driven and Model-Based Fault Classification For Gas Turbines Diagnosis, *Proceedings of ASME Turbo-Expo 2010, Paper No. GT2010-23075,* Glasgow- Scotland, UK, pp1-2.

Maria, A. (1997): Introduction to Modeling and simulation, *Proceedings of the 1997 Winter Simulation Conference* (ed. S. Androdothi, K. J. Healy, D. H. Withers and B. L. Nelson), pp 7-9.

Ogbonnaya E. A., (2004a): *Modeling Vibration – Based Faults in Rotor Shaft of a Gas Turbine,* Ph.D Thesis, Dept. of Mar. Engrg., RSUST, Nkpolu Port Harcourt, Nigeria. pp 82-160.

Ogbonnaya, E. A., (2004b): *Thermodynamics of Steam and Gas Turbines,* 1st edition, Oru's Press Ltd, Port Harcourt. pp 4-5.

Ogbonnaya, E. A., and Koumako, K.E.E., (2006): *Basic Automatic Control,* 1st edition, King Jovic Int'l. Publisher Port Harcourt. Pp 114-115.

Ogbonnaya E.A., (1998): *Condition Monitoring of a Diesel Engine for Electricity Generation,* M-Tech. Thesis, Dept. of Mar. Engrg. RSUST, Port Harcourt, Nigeria. pp 42-43.

Ogbonnaya, E.A., (2009): Diagnosing and Prognosing Gas Turbine Rotor Shaft Faults Using "The MICE", *Proceedings of ASME Turbo Expo, GT 2009-59450,* Orlando, Florida, USA . pp 1-6.

Ogbonnaya E.A and Theophilus-Johnson, K(2010) Use of Multiple Variable Mathmatical Method for Effective Condition Monitoring of Gas Turbines, *Proceedings of ASME Turbo- Expo GT 2010-22568*,Glasgow, Scotland, June 14-18. 2010,

Ogbonnaya, E.A., Theophilus-Johnson, K.,Ugwu, H.U and Orji, C.U(2010) Component Model-Based Condition Monitoring of a Gas Turbine, *ARPN Journal of Engineering and Applied Sciences*, Vol.5, No.3 March. Available at: *www.arpnjournal.com*.

Ogbonnaya E.A and Theophilus-Johnson, K(2011) Optimizing Gas Turbine Rotor Shaft fault Detection, Identification and Analysis for Effective Condition Monitoring, *Journal of Emerging Trends in Engineering and Applied Sciences (JETEAS)* 2(1),11-17 Copright Scholarlink Research Instite Journals (ISSN:2141-7016). Available at:. *http://www.jeteas.scholarlinkresearch.org* and *http://www.scholarlinkresearch.org*.

Pussey, H. C., (2007): Turbo machinery Condition Monitoring and Failure Prognosis, Shock and Vibration Information Analysis Centre/Hi-Test Laboratories, *Proceedings of Institute of Vibration*, Winchester, Virginia.. pp 2-10.

Rieger, N.F., McCosky, T.H and Davey, R.P(1990) The High Cost of Failure of Rotating Equipment, *Proceedings of the 44th Conference of Machinery Failure Prevention Group (MFPG)*, *Vibration Institute*, pp.2-3.

Roemer, M. J. and Kacprzynski, G.J(2000) Advanced Diagnosyic and Prognostic for Gas Turbine Risk Assessment, *Proceedings of ASME Turbo Expo GT2000, gt 2000-30*, Germany. p.10.

Romesis, C and Mathioudakis, K. (2003) Setting up of a Probabilistic Neural Network for Senor Fault Detection Including Operation with Component Fault, *Journal Of Engineering for Gas Turbines and Power*, 125, pp.634-641.

Schneider, E.., Demircioglu, S.; Franco, S., and Therkorn, D., (2009): Analysis of Compressor On-Line Washing to Optimize Gas Turbine Power Plant Performance, *Proceedings of ASME Turbo Expo 2009, GT 2009-59356*, Orlando, Florida, USA. pp 1-4.

Stamatis, A., Mathioudakis, K., Berios, G and Papailiou, K(1991) Jet Engine Fault Detection with Discrete Operating Points Using Gas Path Analysis, *Journal of Propulsion and Power*, Vol.7, No.6, pp.2-3.

Stamatis, A., Mathioudakis, K., Ruis, J and Curnock, B (2001) Real-Time Engine Model Implementation for Adaptive Control and Performance Monitoring of Large Turbo-fans, *ASME 2001-GT-362*. Available at: *http://www.ase.aec.nasa.gov/projects/ishem/paper/vdponi_ac_prop.doc*.

Uhumnwangho, R.; Ofodu, J.C., and Emiri, U. V., (2003): Performance Evaluation of a Gas Turbine Engine, Univ. of Port Harcourt, Nigeria. *Nigerian Journal of Engineering Research and Development*, Vol. 2, No.1. pp 9-20.

Urban, L.A and Volponi, A.J (1992) Mathematical Methods Of Relative Engine Performance Diagnostics, SAE Transactions, Vol. 101, *Journal of Aerospace*, Technical paper 922048. pp.4-5.

Volponi, A.J (1994) Sensor Error Compensation in Engine Performance Diagnostics, ASME Paper, GT1994-58.

Volponi, A.J.,Depold, H. And Ganguli, R.(2003) The Use of Kalman Filter and Neural Network Methodologies in Gas Turbine Performance Diagnostics: A Comparative Study, *Journal of Engineering for Gas Turbines and Power, Vol. 125, Issue 4, pp917-924.*

8

Application of Blade-Tip Sensors to Blade-Vibration Monitoring in Gas Turbines

Ryszard Szczepanik, Radosław Przysowa, Jarosław Spychała,
Edward Rokicki, Krzysztof Kaźmierczak and Paweł Majewski
Instytut Techniczny Wojsk Lotniczych (ITWL, Air Force Institute of Technology)
Poland

1. Introduction

Non-contact blade vibration measurement in turbomachinery is performed during development phase to verify design quality of bladed disk and its structural integrity (Zielinski & Ziller, 2005). The method, referred as blade tip-timing (BTT) or Non-contact Stress Measurement System (NSMS) is applied by mostly all manufacturers as a complement of strain gauges, traditionally used to measure stress levels and blade vibration parameters (Roberts, 2007). Measurement results are usually presented in the function of rotational speed in Campbell diagram, showing vibration modes excited by particular engine orders. Operational stress levels and accumulated fatigue cycles should not exceed material endurance limits.

High Cycle Fatigue, occurring at low stress and high vibration frequency is a common reason for blade damage in turbomachinery. HCF has been identified as factor limiting development of more efficient blade designs, affecting safe operation of turbomachinery and causing considerable losses (Nicholas, 2006). US Air Forces initiated HCF Science and technology program in late 1990's, which launched and supported multidisciplinary efforts for HCF mitigation, continued recently as Engine Prognosis Program. Development of tip-timing instrumentation both for supporting design of fatigue-resistant components and also for online blade crack detection has been one of research priorities and provided new sensors and advanced data analysis methods. NSMS technologies are also developed and successfully applied in power industry (Ross, 2007).

Nowadays Blade Tip-Timing using optical sensors is considered as mature technology able to replace strain gauges in development process of fans or compressors (Rushard, 2010; Courtney, 2011). Current research activities concentrate on turbines, which are more demanding environment for tip-timing instrumentation due to high temperature, contamination and lower amplitude of vibration. Development of alternative tip sensors is considered as another priority. Optical sensors despite providing the highest available resolution, require cleaning and ensure quite limited life, which makes them unusable in embedded systems for blade health monitoring.

This chapter describes development and application of inductive, eddy-current and microwave tip-timing sensors for gas-turbine blades, carried out in ITWL in last five years. Other sensors' applications, like measurement of tip-clearance, blade twist and disk vibration are also discussed here. The sensors' performance has been verified during engine bench tests and in spin facility. They has been used in prototype onboard system for health monitoring of turbofan engine (Przysowa, Spychała, 2008). Similar research efforts have been reported lately by some other leading research centers (Madden, 2010; Hayes et al. 2011; Millar, 2011), revealing not many details. Tip-timing technology is still sensitive area for involved companies and very few journal papers and books are available. State of the art is presented on technical conferences and in patent applications.

More and more interest in blade health monitoring is observed recently, as it allows to maintain legacy turbines, which blades were designed before implementation of CAD systems. It is usually safe and efficient to start monitoring poorly designed blades, instead of canceling or redesigning the machine. In aero-engines fan blades vibration monitoring is performed also to early detect foreign object damage (Gilboy et al. 2009) and to mitigate FOD-initiated HCF cracks.

Online health monitoring may be a method for damage-tolerance maintenance (ENSIP 2004) also for modern blade designs, ensuring turbine operation up to material endurance limits.

This chapter is devoted to problems including sensor design, blade pulse generation and triggering, time-of-arrival measurement, which form only low-level but fundamental part of NSMS. They are common for all applications of BTT, like health monitoring or component testing for design verification or engine certification. Details of TOA data processing, crucial for estimating stress or identification of vibration mode parameters are only mentioned here and are further discussed by other authors (Heath, 2000; Zielinski & Ziller, 2005; Witos, 2007; Loftus, 2010).

2. Parameters of signals received from inductive sensors

Inductive sensors (known also as speed pickups or Variable Reluctance sensors) were primarily used for blade tip-timing in steam turbines (Zablockij et al., 1977). They were successfully employed in monitoring system for blade vibration, which was developed in Poland for turbojet training aircrafts (Szczepanik & Witoś, 1998; Witoś & Szczepanik, 2005). It is the first airborne tip-timing system implemented worldwide and is still in use by Polish Air Forces.

Inductive sensors are optimal for embedded systems due to simple design, low cost and very long life (even several years). The only disadvantage is ability to sense only ferromagnetic materials, which are rarely used in modern gas turbines.

Authors' efforts focused on magnetic field modeling, optimizing amplifier and coil design and selection of strong permanent magnets resulted in development of the sensors usable for non-magnetic metal blades (titan, nickel based super-alloys etc.). Sensors of this type are often called "passive eddy-current". Temperature resistant designs of the inductive sensor for blade tip-timing in gas-turbines have also been demonstrated (Przysowa, Spychała, 2008), able to operate in contact with exhaust gases of temperature even as high as 1200 K. The following subchapters (2-6) review features and applications of advanced inductive sensors, aimed at their implementation in onboard monitoring system.

The signal received from an inductive sensor installed in the area of blade rotation and measured at the output of the conditioning circuit may be like the waveform shown in Fig. 1.

The acquired waveform makes it possible to obtain, after processing the curve by electronic circuits, a number of parameters that can be used for non-contact measurements of blade vibration and momentary tip clearance, i.e.:

tp(n) – arrival time of a blade, counted from the beginning of measurements,

TCx(n) – peak-to-peak amplitude of the x signal,

TCy(n) – peak-to-peak amplitude of the y signal (applicable for two-coil sensors),

tz(n) –falling time for the pulse edge of the x signal,

Ax(n) – peak amplitude of the x signal at the moments of $tp(n)$,

TTn – arrival time for a subsequent blade,

tip clearance.

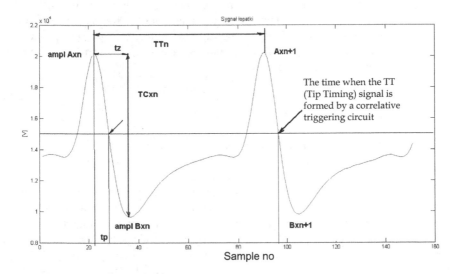

Fig. 1. The waveform of the signal acquired from the sensor with indication of diagnostic parameters

Arrival times serve as the basis for further analysis and make it possible to estimate parameters of blade vibration (Heath, 2000), which can be used for determining condition of blades. However, the signal generated by inductive sensors transfers also other information like tip clearance, rotor unbalance, the angle of blade tip.

3. Non-contact measurement of turbine blade vibration

Measurements of blade vibration (tip timing) are carried out by direct measurements of time intervals between the events when subsequent blades appear below the sensor (Witoś, 1999). Correct and accurate detection of these events is the basis for further processing of measurement signals (Żółtowski, 1984). For non-ferromagnetic blades, the blade is just below the sensor at the moment of the maximum slope for the basic signal. It corresponds to the maximum of the first derivative or the zeroed value of the second derivative.

The sequence of pulse waveforms produced by the sensor is the sensor response to a sequence of events when blades cross the area just below the sensor. That is why we speak about the *event time* having in mind the initial moment when formation of the sensor response is started. The primary carrier of information about the *event time* is always the signal produced by the sensor or, in more details, the information is contained in the invariable parameters of the waveform shape. Then the signal is transferred to the triggering circuit via appropriate signal conditioning modules that provide the most favourable conditions for extraction of required parameters. The nature of the conditioning process and type of the conditioning technology applied depend on the type of a specific sensor and the collaborating blade. To minimize indeterminacy of the moment when the signal level exceeds the reference threshold, the maximum possible slope of the signal waveform is required at the detection point with the minimum possible noise dispersion. It is the requirement that needs maximization of the signal slope with respect to the noise and is referred to as the acronym (*slope-to-noise ratio*).

$$SLNR \overset{def}{=} \frac{[\frac{dV_i(t)}{dt}]_{T_K}}{\sigma_N} \tag{1}$$

The task of the triggering circuit that is meant for extraction of time-related information is to produce a standard logic pulse (so-called Tip Timing – TT) with the edge that remains in a defined, fixed relationship to the time coordinate of the event. To detect the moment when a blade appears below the sensor one can use threshold discriminators or comparators. Depending on the adopted measurement methods, these devices are operated in a conventional way, with the non-zero reference level or as zero crossing detectors. The produced logic pulse (the TT signal) is delayed against the event of a blade arrival. Thus, the time necessary to produce the pulse shall be referred to as the instrumental time. For single-channel measurement, the delay is not very important provided that it maintains its fixed duration, in spite of waveform variations for the sensor signal. In case of multi-channel measurements, the instrumental time may vary, despite identical configuration of electronic components in the channel. That inequality may be caused by parameter dispersion of components of the amplifier or the sensor.

Fig. 2 and Fig. 3 present waveforms of turbine blade vibration recorded during ground tests of a turbojet.

The estimated vibration of turbine blades feature the low amplitude at the level of 0.1mm, but application of the constant fraction detection made it possible to achieve a very high resolution that enabled measurements of vibrations with the accuracy as high as up to 0.01%.

Fig. 4 shows the normalized amplitude of resonance vibrations for a single turbine blade. The relationship presented in Fig. 4 serves as the basis to plot the Campbell diagram. It allows to determine the excitation orders for the vibration synchronous with the rotational speed. The observed vibration in the first bending mode is excited by the fifth and higher engine orders (Fig. 5).

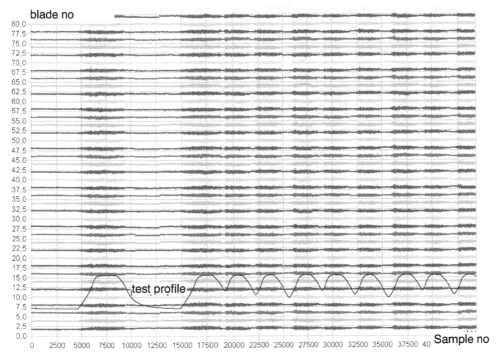

Fig. 2. Blade vibration amplitude and rotor speed in subsequent rotations of the turbine

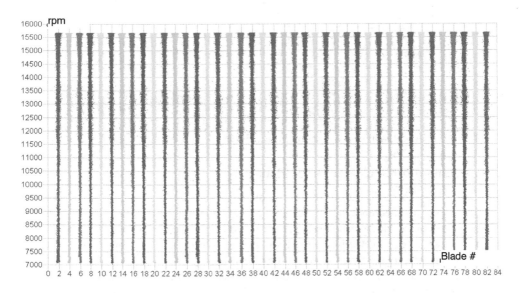

Fig. 3. The relation between blade vibration amplitude and rotor speed

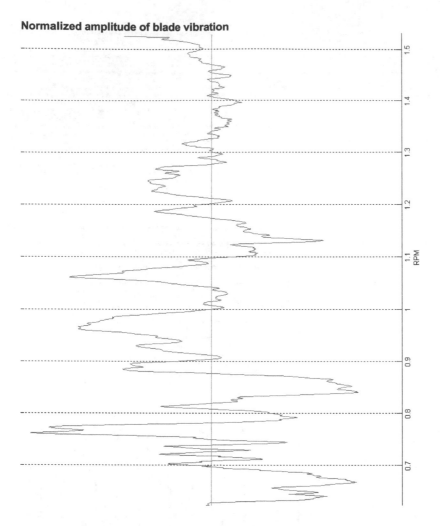

Fig. 4. The relation between the vibration amplitude of a selected blade and rotor speed

4. Measurement of tip clearance for blades

Additionally, it was evaluated whether the applied measurement chain is suitable for determining the tip clearance above turbine blades rotating in turbojet engine. The tip clearance of a selected blade measured with an inductive sensor for the entire range of rotor speed is shown in Fig. 6. Design properties of that sensor are different from the ones demonstrated by the sensor that is used for the tip timing method.

During rotor acceleration and deceleration the measurement results indicate clearance variations characteristic for turbomachinery (Fig. 7) caused by different parameters of thermal expansion for blading and casing.

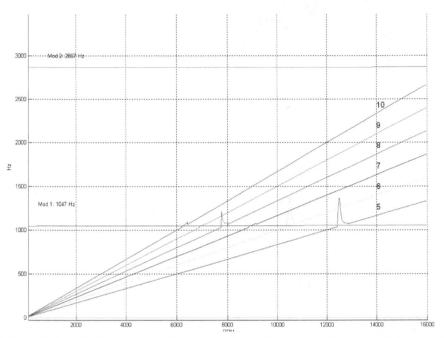

Fig. 5. The relationship between the frequency of a turbine blade vibration and the rotation speed – the Campbell diagram

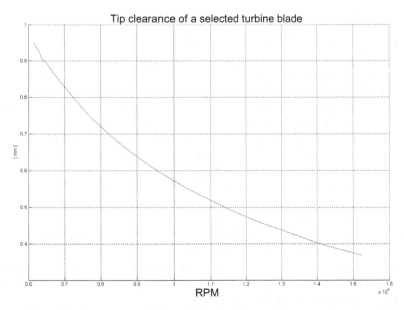

Fig. 6. The relationship between variations of the turbine rotation speed and variations of the tip clearance

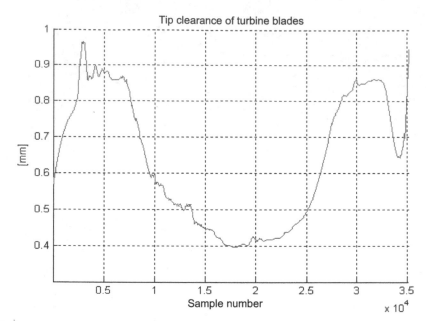

Fig. 7. Average tip clearance of turbine blades for an example test profile

5. Measurements of twist angles for blades

The section 2 reports the possibility of determination of additional diagnostic parameters on the basis of the measured signals. For that purpose a software detector of the signal peak values was developed by means of procedures embedded in the Matlab software tools. The peak detector was provided with components of the constant fraction detection of signals, which improves insensitivity of the determined parameters to amplitude variations of the signal obtained from the inductive sensor. The signal from the peak detector served as the basis for determination of arrival moments of blades. In turn, the knowledge of arrival times makes it possible to calculate twist angles of blades at the moments of arrival.

It should be mentioned that the concept of measurement of angles between blades and the measurement plane is based on the formula for mutual induction (Rokicki et al. 2009). The initial laboratory tests were carried out with use of a sensor with two coils. At first the disk with blades was prepared in such a way that the disk has 1 titan blade and two blades with different twist angles (Fig. 8). Then the sensor signal was recorded (Fig. 9) for three different angles of the sensor position (0°, 45° and 90°).

Fig. 9 and 10 present signals for various twist angles of blades. Low amplitude and half-round pulses are the sensor response to arrival of a titan blade.

The amplitudes of pulses collected from both coils of the sensor as well as arrival angles for individual blades are shown in Fig. 11. Therefore, the possibility of determination of angular positions (twist angles) for blades was confirmed.

The newly developed methodical approach and algorithms were subsequently used to find out arrival angles of blades during turbojet bench test (Fig. 12).

Signal amplitude

Fig. 8. View of the turbine disc with blades

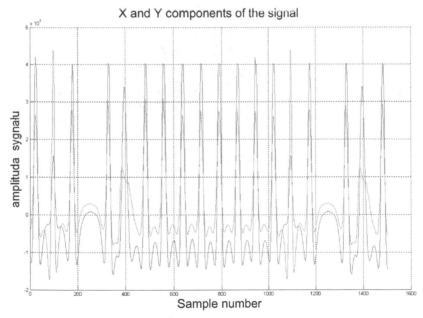

Fig. 9. Signals from the two coils of the sensor (the setting angle of the sensor is 0º).
X component waveform is green and the Y component is blue.

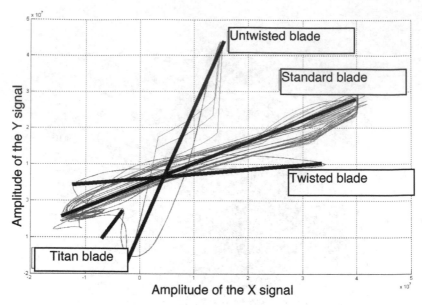

Fig. 10. The graph for arrival angles of blades of the disk as shown in Fig. 8

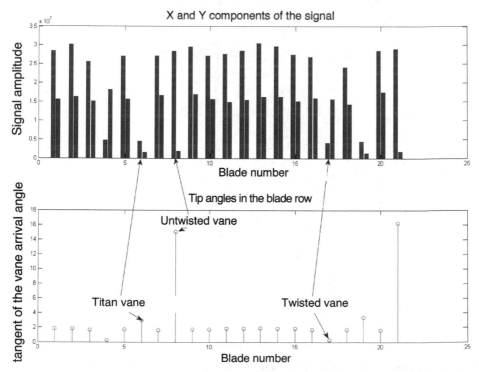

Fig. 11. Component amplitudes of signals and the blade arrival angle

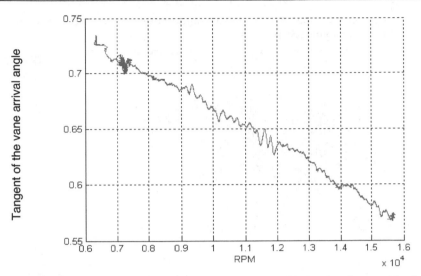

Fig. 12. Calculated tangent of the arrival angle for a selected blade of turbojet as the function of the rotational speed

6. Estimation of rotor unbalance

The following paragraphs comprise results of concept developments, where the signal received from the inductive sensor would be used for measuring the unbalance.

Two inductive sensors with the transverse field were installed in the rotation area of blades with the initial clearance of 1 mm. Then, a series of tests was carried out when controlled unbalance was introduced to the 7th stage of the compressor. The signals received from both sensors were conditioned with use of the same circuits and then recorded (Fig. 13 and Fig. 14). One of the sensors was dedicated for the tip timing method whilst the second one was

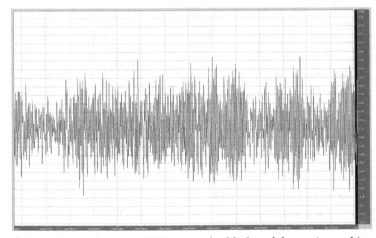

Fig. 13. Signal from the tip clearance (TC) sensor for blades of the engine turbine when the setting for the tip clearance at rest was 1 mm.

meant to measure tip clearance. Measurements of the turbine unbalance were carried out by means of the signal from the second sensor.

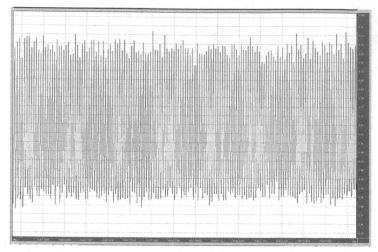

Fig. 14. Signal from the tip timing (TT) sensor for blades of the engine turbine when the setting for the tip clearance at rest was 1 mm.

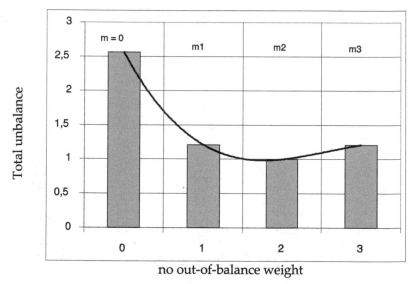

Fig. 15. Effect of the compressor out-of-balance onto total out-of-balance of the turbine

The self-developed algorithms implemented within the Matlab software tool were used to find out how the compressor out-of-balance affects the total out-of-balance of the turbine (Fig. 15). Before tests the compressor was balanced with use of the Schenck balancer. The best balance was achieved when no additional weight was used. Then the appearing out-of-balance of the compressor was measured for the other weights (m1 < m2 < m3).

On the basis of the tests one may conclude that the compressor out-of-balance increases in pace with the growth of the add-on weight. The completed measurements made it possible to find out that for the best balance of the compressor (no additional weight) the turbine suffers from nearly three times higher vibrations than for all the other cases. The effect of insufficient load to the bearings onto the turbine operation is manifested here. For the largest weight $m3$, the maximum level of vibrations permitted for that type of engine was not exceeded.

7. Application of microwave sensors

Following subchapters present prototype microwave sensors that are developed in the Air Force Institute of Technology (ITWL) as well as results and conclusions obtained from their testing. These devices represent radars of a very short range, needing special designs of antennas and dedicated methods for signal processing. Particular attention is paid to analysis of the way in which the engine environment and the selected sensor design affect waveforms and parameters of the signal acquired from turbine blades.

The major advantage of microwave sensors is their resistance to contamination by combustion products, relatively good propagation of microwave radiation, both in wet air and in the environment of hot exhaust gases and good reflection of the microwaves by metals. On the other hand, the drawbacks include (Nyfors, 2000):

expensive manufacturing process of the microwave devices, in particular those that are meant to operate at high temperatures and frequencies (above 12 GHz),

the need to tune operation frequency to match dimensions of the equipment and the resulting infeasibility to develop a universal design,

the sensor signal is usually a function of several variables, therefore the sensors need calibration, advanced processing of signals or application of a reference sensor,

limited spatial resolution resulting from quite a considerable wavelength (as compared to optical sensors).

Microwave sensors were already applied to turbine engines for various applications, i.e. to measure the content of wear products in oil inside the gearbox or to monitor operation of the fuel pump (Błachnio et al., 1985).

Microwave sensors intended for measurements of motion parameters of turbine blades have several common features with capacitance sensors based on frequency modulation (Forgale, 2006) and may be considered as an alternative solution.

8. Microwave sensors developed by the Air Force Institute of Technology (ITWL)

Research studies have led to the development of several design solutions dedicated to microwave sensors of blade motion. All the designs are based on the concept of a homodyne detection of microwaves reflected on blade tips. Such a concept was for the first time applied to MUH sensors (Dzięcioł, 2003, 2004).

The microwave homodyne device (MUH - Fig. 16) is made up of a steel and ceramic antenna that is resistant to high pressures and temperatures as well as a generator and detector. The sensor is installed on an engine by means of the antenna screwed onto the socket provided on the engine body. The blades that pass in front of the sensor trip cause retuning of the resonance circuit of the antenna. Consequently, the sensor produces the measurable signal that is the fast-varying component of the output voltage U provided by the microwave detector.

The design that is more suitable for tests of engines is the microwave sensor in the form of a compact homodyne generator and detector (GDH) that can be screwed onto the existing antenna (Fig. 17).

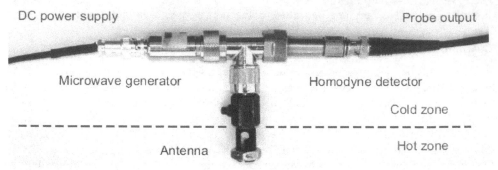

Fig. 16. The MUH microwave motion sensor – the laboratory version made up of coaxial subassemblies connected by means of a tee piece with N-type connectors

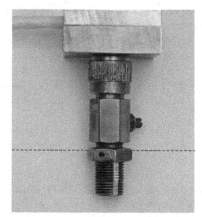

Fig. 17. The MUH microwave motion sensor (the version with the GDH module)

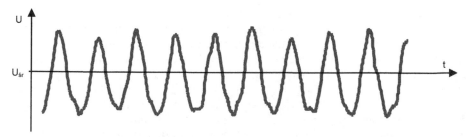

Fig. 18. Model waveforms for the voltage signal acquired from turbine blades with use of the MUH sensor

The following research efforts resulted in the design of a differential antenna (Fig. 19, Rokicki et al., 2007) with the aperture that is made up of two open waveguides, which interact to generate the output signal.

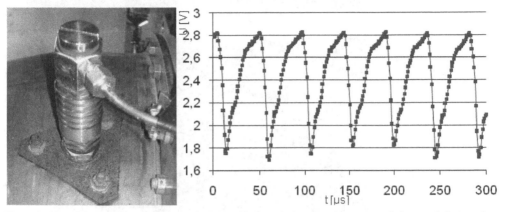

Fig. 19. The ITWL differential antenna installed in the turbine casing and the model signal waveform acquired at the maximum engine speed

9. Overview of exiting solutions for microwave sensors

The issue concerning the ways of employing microwaves for measurements of the blade motion was investigated by several research centres, mostly located in the USA. The device developed by United Technology Corporation (Grzybowski, 1998) operates with the frequency of 20 GHz and is intended for tip-clearance measurement. The sensor from Daimler-Benz (Wegner, 1997) is also dedicated to measure tip clearance and operates with the frequency of 22-24 GHz. It is made up of a metallic and ceramic antenna and an electronic module incorporating microwave monolithic integrated circuits (MMIC) based of gallium arsenide (GaAs).

The device of Siemens (Wagner, 1998) is intended for monitoring of vibrations of steam turbine blades, The generator operates at the frequency of about 24GHz that can be adjusted by its control circuit according to the actual distance to the blades. Pulses generated by passing blades are initially processed by the DSP circuit and then analyzed by a computer with the appropriate software for the discrete phase method.

The solution from Radatec (Geisheimer et al., 2002; Holst, 2005) ($f = 5.8$ GHz) is distinguished by application of two detectors with the mutual phase shift of 90°, which makes it possible to measure both the in-phase and the quadrature components of the wave vector. Comparison between the both signals makes it possible to unambiguously measure the tip clearance by elimination of the effect of received microwaves on the amplitude.

The EHDUR device of BAE Systems is dedicated to detect foreign bodies in the engine inlet and to identify the objects that may damage the engine. Initially the motion of blades was considered by the designers as disturbance and filtered out of the input signal (Shephard et al., 2000). More recently the studies were initiated to analyze blade vibration of fan blades on the basis of the signal generated by a sensor installed at the inlet.

The prototype device developed at the Industrial Telecommunication Institute (PIT) (Fig. 20) developed in collaboration with ITWL offers excellent parameters of both the generator and the antenna circuits. The antenna and generator /detector modules are detached and connected by means of a semi-rigid cable.

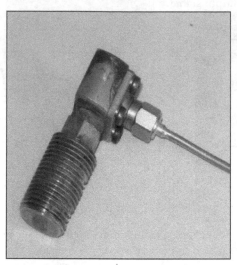

Fig. 20. The antenna for the PIT microwave sensor

10. Operation principle of the microwave sensor

Non-contact measurement of blade vibration requires the sensor providing electric pulses in the moments when blades cross the measuring point of the antenna (Zablockij et al., 1977). Microwaves, emitted by the antenna, return to the circuit after having been reflected by tips of rotating turbine blades. The resulting signal, obtained as a result of the phase-sensitive detection of microwaves, represents the momentary location of blades (Fig. 21). The way in which the signal from the MUH sensor is generated, is outlined in the thesis (Dzięcioł, 2003). Under real working conditions of the turbine, operation of the microwave circuit is affected by a number of factors that are hard to describe in quantitative terms. The detailed analysis of the sensor operation principle is not within the scope of this study, but understanding of the specific features is necessary to analyze the received signals.

Modelling of the sensor operation requires the solution of the Maxwell equation, which is carried out with use of numerical methods by means of the dedicated software. It is really difficult to achieve exact results of simulation due to the fact that some parameters that are necessary for calculations cannot be determined with sufficient accuracy, especially, when operation of the sensor at high temperatures, typical for gas turbines, is anticipated.

Specific operation of the sensor during engine tests can be analysed by means of simple models that are known from physics and electrical engineering, such as a transmission line, a dipole, interference of waves, etc., Application of such models is limited due to the fact that dimensions of the circuit and the measured distances are comparable with the wavelength or they are even shorter. The optimum method for development of the microwave sensor for turbine environment must incorporate intense aid of the computer simulation methods to the experiments.

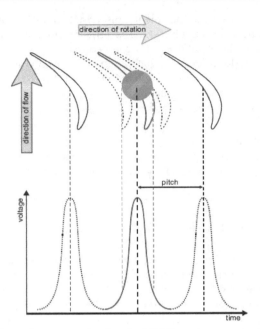

Fig. 21. The principle for pulse generation by the MUH motion sensor. For antennas with low spatial resolution with the range that covers signals from 2 – 3 blades the pulses may interfere.

The structure of the employed microwave circuit is made up of two major parts:
- closed – waveguides that are included inside the sensor,
- open – turbine gas path with blades rotating inside.

The closed part (except for the sensor antenna) is located in the cold area that enables steady operation of the incorporated components, as the temperature alters their parameters only slightly. On the other hand, electromagnetic parameters of exhaust gases that fill the turbine channel may vary depending on locations and changes of the engine operation range.

10.1 Operation principle of the antenna

To enable the radiation to reach the blades, the steel-ceramic antenna must perform as the component that matches the microwave circuit to the space when motion of the rotating parts takes place. Due to the temperature effect onto the antenna, its actual parameters, when it is mounted on the engine, are different from calculated ones. The level of usable signal is lower for higher rotation speed values, as a significant deal of the generator power is already reflected and cannot be emitted.

Microwaves are reflected not only by blade tips, but also by other metal components. It is difficult to limit the sensitivity of conventional antennas exclusively to the objects of selected type (shape). Generally, it is only possible to set up the antenna polarity and provide the directional characteristics.

In the case of the sensor intended for monitoring the motion of blades, the distance to the object (tip clearance) is low as compared to the wavelength. The examined object is no longer passive with respect to the antenna, but begins to interact with it. In the case of the

MUH sensor, the blade is a part of the antenna resonance circuit and arriving of the blade to the antenna field retunes the circuit, which is reflected in the signal received from the detector (Dzięcioł, 2003). The effect was also observed in the tested PIT sensor.

The differential antenna (Rokicki et al., 2007) emits nearly no waves into the surrounding space, but reacts to the presence of objects right in front of it (1-2 mm). Consequently, a very high spatial resolution is achieved, which is confirmed by the signal with a very short fall time and low duty factor of the pulses.

10.2 Amplitude of pulses

The sensor employs the homodyne (synchronous) detection of microwaves, therefore the output voltage U_{wy} depends on the phase shift φ between the received and reference signals:

$$U_{wy} = a\,A \cos \varphi, \tag{2}$$

where:

A – amplitide of the received wave,

a – proportionaliy constant.

As the wavelength is about 3 cm, the differences in the tip clearance for individual waves (typically less than 1 mm) result in only insignificant changes of the phase, which can be observed in the received signal. In fact, besides the phase relationships, amplitude of the obtained pulses is also affected by the power of the received signal (Dzięcioł, 2003). The pulse amplitude not only depends on the tip clearance, but also on the cross section area of the blade and on local phenomena that may attenuate propagation of microwaves. Exact separation of all these factors is impossible in the existing circuit.

Variations in the distance between the antenna and the blades (tip clearance) due to changes of the engine operation range are insignificant as compared to the wavelength and should not significantly attenuate the signal level. When the tip clearance is altered by 0.5mm, the distance covered by the wave is doubled:

$$\Delta x = 2 \cdot 0,5 \text{ mm} = 1 \text{ mm} \tag{3}$$

From the equation for the stationary wave, the following can be calculated (under the assumption that the wave length is $\lambda = 3$ cm, just as in the air):

$$A = \cos(2\pi\,\Delta x / \lambda) = 0,978 \tag{4}$$

The voltage output signal from the microwave detector (after elimination of the constant component) is proportional to the amplitude A of the received wave, so the relative alteration of the output signal is:

$$\Delta U = 2,2 \text{ \%} \tag{5}$$

For real turbine channels the wavelength may vary from one location to another or in time, in pace with the electric permittivity of gases due to pressure and temperature variations.

10.3 Wave effects

Propagation of radiation takes place in the field near the antenna, so wave effects, such as diffraction, interference, the Doppler effect, as well as reflections from other objects, not only from blade tips, substantially influence the received signal. All the effects deteriorate operational parameters of the sensor, chiefly the spatial resolution and the level of usable signal.

In the case of the simplest antennas (open waveguides) one can assume that points that belong to the antenna aperture can be considered as sources of spherical waves (Fig. 22). The field that is generated in the channel of an immobile turbine constitutes the system of stationary waves that resembles interference images. The resulting signal that is received by the antenna is the superposition of signals reflected on blade tips, but also on vanes, blade side surfaces or disk surfaces, in particular when blades are short and densely spaced.

The wave that is reflected on the area between the blades (unless it comes with the phase opposite to the generator) is also received and must be considered as the parasite component of the usable signal – increases the background level and reduces the signal/noise ratio of blade pulses. To avoid undesired reflections the antenna design must guarantee a possible short range, not much exceeding the values of the tip clearance (below 5 mm). The aperture diameter should be comparable with the thickness of the blade leaf and much shorter than the distance between two neighbouring blades. In practice, it is very difficult to meet all these requirements. Development of a miniature and selective antenna requires different design approach and the application of untypical components and expensive technologies.

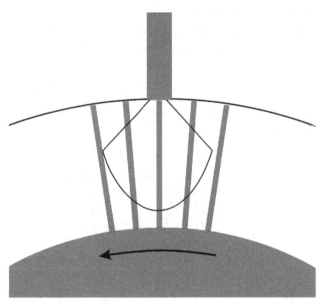

Fig. 22. The antenna with excessively large range. The area is marked where reflections may possibly interfere with the output signal of the sensor. The proportions are maintained between the antenna diameter on one side and dimensions and layout of turbine blades on the other side.

Influence of the Doppler effect can be neglected under the assumption that blades move perpendicularly to the probing wave. In practice that assumption may not be fulfilled and the effect of the blade velocity onto the output signal from the sensor is observed. The lack of possibilities of separating the effects of various factors and physical phenomena (location, speed, attenuation) on the output signal of the device restricts its usefulness and applications.

11. Processing of signals received from blades

11.1 Introduction

All measuring systems that have been in use so far employ magnetic or capacitance sensors combined with analog triggering circuits. Such systems, even after updating, are incapable of providing the appropriate values of parameters, necessary to measure vibrations of gas turbine blades. Signals from dedicated microwave sensors as well as from capacitance ones with frequency modulation can guarantee that the desired resolution and reliability are achievable. Due to the shapes of received pulses, such sensors require application of dedicated methods for signal processing. In this study the method that enables determination of amplitudes and frequencies of turbine blade vibrations as well as the tip clearance by analysis of the digital signal received from sensors is proposed.

The newly developed algorithms, in spite of the fact that they are chiefly dedicated to signals from microwave sensors, were also successfully tested for signals received from compressor blades by inductive and eddy current sensors.

12. The concept of signal processing

Information about movements of individual blades is obtained in the form of a series of samples that make up the blade timings with resolution (quantization) defined by parameters of the analog-to-digital (A/D) converter (Fig. 23). Shapes of the received pulses depend on the sensor type and geometrical parameters of the blade row. Tip deflection is reproduced for each blade in the measurement waveform received from the microwave sensor, primarily by shifting the pulse down the time scale with regard to the expected location. Extraction of that information from the sampled waveforms with the required resolution makes it possible to measure blade vibration.

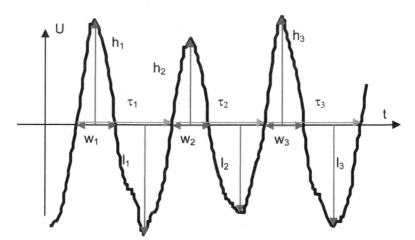

Fig. 23. Parameters of subsequent pulses acquired from turbine blades with use of a microwave sensor: amplitudes of peaks h_i and background level l_i (with regard to the reference voltage), pulse width w_i and time gaps τ_i between the moments of signal reception from subsequent blades.

To determine momentary deflections of the selected blade, it is necessary to unambiguously identify fragments that are associated solely with the blade movements and to extract that information from the sampled waveform in order to obtain the corresponding pulses. Then these pulses can be described by means of such parameters as the amplitude, phase and width (Fig. 23) that are measured against the selected reference voltage. After these values are determined for all pulses and assigned to specific blades, it is possible to convert the measurement waveform (in the form of samples) into a sequence of ordered matrixes of real numbers that correspond to location of blades (radial and angular coordinates) for subsequent revolutions (Fig. 24). These numbers serve as the input data for further vibrations analysis of blades and the rotor unit.

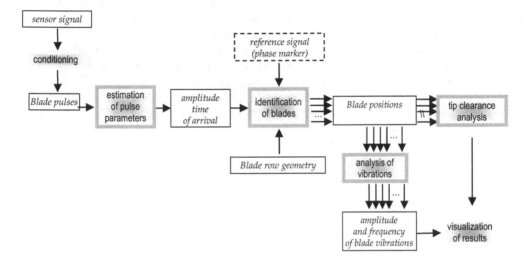

Fig. 24. Procedure for processing of signals acquired from blades

13. Conditioning and recording of signals

The voltage signal from a microwave sensor has usually a constant component that corresponds to the working point of the detector. That constant component is usually much higher than the fast-changing signal received from blades. Prior to amplification of the output signal, it is necessary to eliminate the constant component or the compensate it. The gain value must be appropriately adjusted so that the dynamic range of the converter would be utilized in the best possible way and signal cropping could be avoided.

To locate pulse peaks with sufficient accuracy the sensor signal must be sampled with high frequency, preferably a dozen times higher than the maximum frequency of blade arriving. This recommendation results from the fact that the pulse peak usually takes less than 30% time assigned to a single blade and at least several samples are necessary to reproduce the peak shape. Only the high oversampling degree of the input signal makes it possible to measure blade vibrations, lower sampling frequencies enable merely to measure the turbine rotation speed.

Results obtained by continuous sampling of the input signal with the constant frequency of f_s were recorded on a hard disk, which was possible owing to relatively high capacities of mass storage memory units of contemporary computers. In the target on-board measurement system the hardware resources must be used economically, thus sampling shall cover only those fragments of pulses that convey information about momentary positions of blades.

14. Measurements of pulse parameters

The basic principle of non-contact measurements of blade vibrations is determination of the moments when subsequent blades arrive to the measuring point of the sensor. The fundamental precondition to carry out measurements correctly is to reproduce such moments accurately by the output signal of the sensor (by the conventional phase of pulses). To determine that phase, it is mandatory to maintain unaltered shapes of pulses provided by the sensor, whilst the pulse amplitude and width may be altered proportionally.

Due to the aerodynamic shape of the blade tip and sophisticated principle of the sensor operation, accurate indication of the point that belongs to the measurement waveform and that corresponds to the conventional moment of the blade arrival (Witoś, 2007) is a really non-trivial task. For inductive sensors, the presence of a blade at the measuring point is associated with transition of the falling edge of the pulse via the voltage level of 0V (Fig. 25). In the case of microwave sensors it is assumed that the presence of a blade at the measuring point of the sensor corresponds to the pulse peak, i.e. the local extreme for the voltage waveform (Fig. 26). Depending on the signal polarization it can be either the signal minimum or maximum.

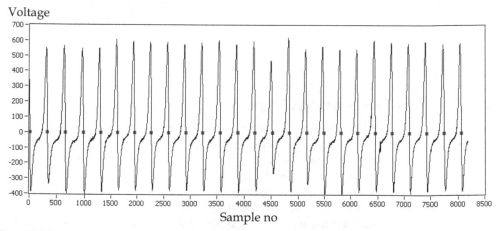

Fig. 25. Measurement of the blade arrival moments with use of digital processing of signals from an inductive sensor. Data acquired from 15 blades of the 3rd compressor stage of a helicopter engine. Localization of transition moments when falling edges of pulses acquired from blades cross the voltage level of 0V (with the hysteresis $\Delta U = 100$ normalized units) was carried out in the domain of sample numbers (the integer variable). Resolution for time measurements is directly associated with the sampling frequency $f_s = 300$ kHz.

The sampled signal received from the sensor is quantized in the domain of amplitude and time. Typical algorithms for processing digital signals are based on samples in the domain of time units for the clock series that synchronizes the sampling process. If no interpolation is applied, the achieved resolution for time intervals is equal to the sampling period.

It is really expensive to increase the sampling frequency up to the value that guarantees that the desired resolution necessary to measure vibrations of blades is achieved, in particular when the sampling results are entirely recorded on a mass storage unit for further analysis. The data acquisition module NI DAQPad 6070E enables sampling with the maximum possible frequency of 1,25 MHz. The achieved resolution in time domain ($\delta t = 0,8\ \mu s$) is many times lower than the one for measurements of digital sequences with use of counters and the dedicated PCB module NI 6602 with the clock frequency of 80 MHz as well as for the SPŁ-2b recorder (10 MHz).

In order to achieve the time resolution exceeding the sampling period δt and avoid quantization, it is necessary to convert a sequence of samples into a continuous representation, at least within the vicinity of local extremes (pulse peaks). The linear interpolation, although much simpler for implementation, poorly reproduces the nature of the sensor waveform. It is why the walking adjustment of a sample clusters was applied, where the clusters included several measurement points and the waveform was approximated with the third degree polynomials with use of the least square method (Fig. 26). The obtained sequence of polynomials appears very similar to the original analog signal.

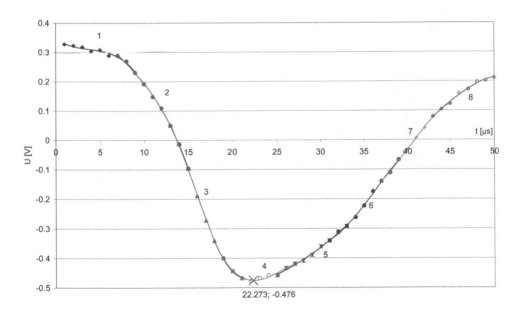

Fig. 26. The polynomial approximation result of a model pulse acquired from blades with use of a microwave sensor. Sampling frequency $f_s = 1$ MHz; window width $w = 9$. Exclusively the derivative of the polynomial no. 4 has zero points, where one of the extremes determines coordinates of the pulse peak, marked with the 'X' sign.

The least square method is commonly used to match the finite set of measurement points to a theoretical curve. The presented procedure analyzes the data one sample after another and calculates coefficients of the third degree polynomials. Such a procedure is beneficial due to the possibility of implementing it in a real time mode. There are other algorithms available that are meant for similar tasks (e.g. spline interpolation), but they all require the preparation of a finite set of data and determination of boundary values with further awaiting until the data processing is completed.

Determination of the polynomial coefficients makes it possible to calculate the first and second derivatives and to check whether the condition of a local extreme is fulfilled. It is possible to speed up the data processing procedure when the analysis is limited exclusively to the points with amplitudes that fall within the predefined limits (threshold voltages).

Accurate determination of arrival times of blades substantially depends on correct selection of the sampling frequency and the width of the window that defines number of samples that are selected for the single approximation. When four samples are selected, the obtained curve must incorporate these points, i.e. the polynomial interpolation shall be applied. When more than four samples are used the applied approximation is a form of the signal filtration. The window width is adjusted to the noise content in the sensor signal (signal to noise ratio) and the number of points used to reproduce the blade waveform.

The outlined method makes it possible to determine location of pulse peaks (Fig. 27) with the resolution that is several times higher than the sampling frequency. Times of blade arrivals obtained with use of the proposed method have the form that is appropriate to determine amplitudes and frequencies of blade vibrations by means of the discrete-phase method (tip-timing) and also its differential version, used when no one-per-revolution signal is available (Witos, 1994).

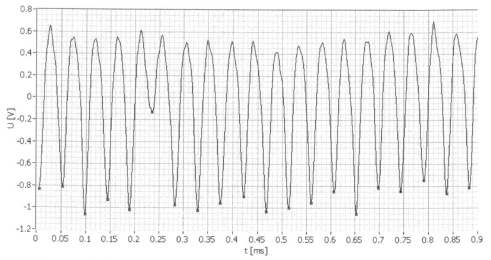

Fig. 27. The example that explains how to determine pulse peaks by means of the newly developed method. The signal from the GDH sensor was used for the rotation speed of $n = 15500$ rpm. One blade generates the pulse with lower amplitude since its tip had been already trimmed.

15. Vibration measurements for turbine blades

Below there is an example of the application of the proposed method to the signal from the ITWL sensor, where the signal was recorded with the sampling frequency of $f_s = 1$ MHz. Then a fragment of the signal recorded for deceleration of the engine (Fig. 28) was subjected to the analysis with use of the discrete-differential method. The results obtained for the spacing error should be considered as trustworthy since the positions of blades are recognized on the basis of waveforms with high fidelity (exceeding 90%).

Blade vibration alter blade spacing by no more than 1-2%. Deviations of the blade spacing in relation to average positions, resulting from non-uniform deployment of blades on the blade row, are of the same range. In order to illustrate blade vibration on the cumulative graph, the spacing error is magnified by several dozen of times with elimination of its constant value (Fig. 31). Consequently, it was found out that the vibration level of blades usually fits within the boundaries $A < 0.4$ mm.

For the rotation speed of n ≈ 8600 rpm the static displacement of blades was observed for the entire blade row. The phenomenon took about a half of a second and affected the amplitude of received pulses (Fig. 32). It resembled synchronous vibrations of blades with the amplitude $2A < 0.6$ mm, but it rather was an effect of a radial or axial movement of the entire turbine. No symptoms of similar turbine behaviour could have been observed for other recorded signals.

n [rpm]

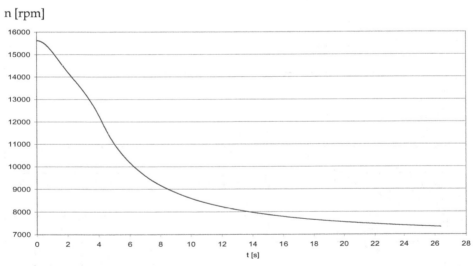

Fig. 28. Recorded deceleration of the tested engine for which vibration measurement of turbine blades is demonstrated.

For most of blades the response of a blade to the input function for the basic form of transverse vibration is visible for the 6th synchronous rotation speed for the rotation speed range from 12,000 to 13,000 rpm. For blades with higher frequencies of free vibrations, the response to the 5th synchronous speed is not visible as it falls outside the range of maximum permissible speed values. The responses for the seventh and eighth synchronous speed have insufficient amplitudes and cannot be identified for all blades. The responses for higher synchronous speeds occur for the rpm values below the idle run.

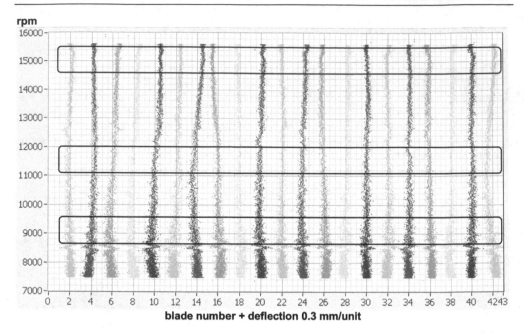

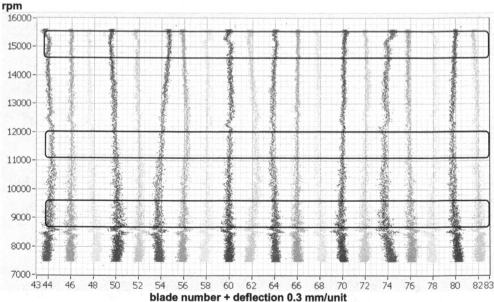

Fig. 29. The relationship between vibration amplitude of turbine blades and rotational speed during the engine deceleration. The image presents differential deflections for even blades

For selected blades the method proposed by (Zablockij et al., 1977) was used to determine frequencies of synchronous vibrations enforced by the 7th, 6th and 5th multiplicity of the rotation speed (Fig. 31, Fig. 32). Then the obtained frequencies were applied to plot the

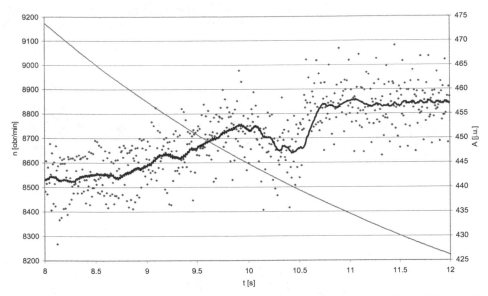

Fig. 30. Amplitude of pulses acquired for the blade no. 1 in the function of time. The graph presents also the trend line (the moving average for the window width of 20) and variation of the rotational speed.

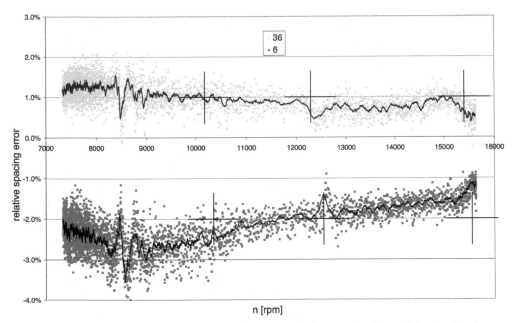

Fig. 31. The relationship between the spacing error and the rotational speed for the blade no. 36 and 6. The image presents the resonance effect for the 7th, 6th and 5th engine order.

Campbell graph (Fig. 33). The achieved relationship between the frequencies of blade vibrations and the rotation speed is quite similar to the results obtained by manufacturer of the blades.

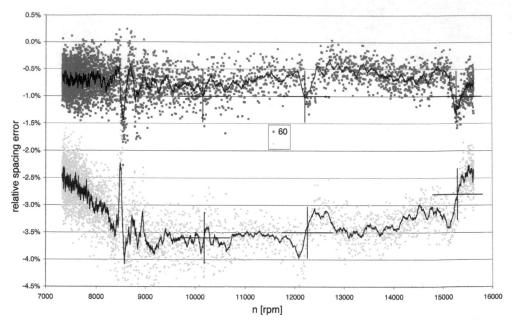

Fig. 32. The relationship between the spacing error and the rotational speed for blade no. 60 and 37. The image presents the resonance effect for the 7th, 6th and 5th engine order.

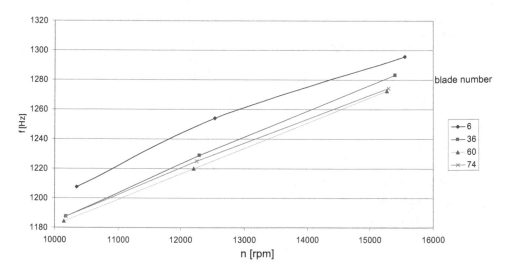

Fig. 33. The relationship between the vibration frequency for selected turbine blades and the rotational speed – Campbell diagram.

The equipment limits for the available instruments (chiefly insufficient spatial resolution and the need to measure very short time intervals $dt < 1$ μs), are the reasons that the results from measurements of the vibration amplitudes are burdened by a wide bandwidth noise. It makes the spectral analysis of the spacing error with use of the FFT method as well as the analysis of asynchronous vibrations difficult. Nevertheless the achieved results make it possible to find out blades with vibration frequencies that fail to meet technical requirements and to monitor decrease of free vibration frequency as a result of fatigue crack development.

16. Conclusions

The high-temperature inductive and microwave sensors along with algorithms for data processing developed by the researchers of the Air Force Institute of Technology (ITWL) enable diagnostics of gas turbines. The sensors are suitable for a long-term operation at temperatures typical for gas turbines and can withstand vibrations that occur during operation of jet engines.

The measurement results show that blade vibrations of a turbine in a jet engine achieve the amplitude near 0.1mm for typical input functions and usually not exceeding 0.4 mm under resonance conditions. For the start-up ranges of the engine the circle sections of such a length correspond to the rotation time less than 1 μs. Therefore, the uncertainty related to the determination of time moments for vibration measurements system dedicated to turbine blades should never exceed the value of several dozens of nanoseconds. On the other hand, not only is the signal processing system the source of deviations, but also the microwave sensor itself. The best spatial resolution was achieved for sensors that produce waveforms with a low duty factor of output pulses.

Application of the newly developed method for signal processing enabled, for the first time, to measure vibrations of turbine blades for the aircraft engine with use of the discrete differential method on the basis of signals acquired from prototype microwave sensors. It required a long series of tests with the engine running until the signal was acquired that met the imposed requirements. Alongside, a numerous numerical experiments with the recorded measurement data had to be carried out.

Anyway, non-contact vibration measurements for turbine blades in an aircraft engine are really difficult due to a very low level of vibration amplitudes and unfavourable working environment for the sensor. The conditioning methods developed, supported by technologically advanced circuits for triggering and data acquisition, made it possible to achieve the accuracy better than 0.01% for the tip deflection measurement. The available resolution is sufficient to estimate blade vibration frequency and identify the vibration mode. Developed methods are capable to warn about the excessive level of blade vibrations as well as about frequency change of the synchronous blade vibrations caused by fatigue cracks.

17. References

Błachnio, J.; Preckajło, S.; Dzięcioł, E. & Sypnik, R. (1985) *A non-contact microwave method for measurements of mechanical motion inside hermetically sealed chambers with flow of stuff under variable pressure* (in Polish), Patent application nr P-259221

Courtney, S. (2011) A Robust Process for the Certification of Rotating Components Using Blade Tip Timing Measurements. *Proceedings of ISA International Instrumentation Symposium,* June 2011, St. Louis, MO USA

Dzięcioł, E. (2003). *A Microwave Homodyne System for Condition Monitoring of a Turbine Blade in an Aircraft engine.* PhD Thesis (in Polish), Air Force Institute of Technology (ITWL), Warsaw

Dzięcioł, E. (2004). *Method of a Continuous Determination of an Instantaneous Position of an Blade Tip in a Rotor Turbine Machine.* US Patent 6,833,793

ENSIP (2004). *Engine Structural Integrity Program.* MIL-HDBK-1783B, Department Of Defense, USA. Available from: www.everyspec.com

Fogale nanotech (2011). Capacitive Blade Tip Clearance & Tip Timing measurement system. In: *http://www.fogale.fr/media/blade/DS_TUR_UK_V02_WEB.pdf*

Geisheimer, J.; Greneker, G. & Billington, S. (2002). *Phase-based sensing system.* US Patent 6,489,917

Gilboy, M.T.; Cardwell, D.N.; Chana, K.S. (2009). Autonomous foreign object impact detection using eddy current sensors. *Proceedings of 4th EVI-GTI International Gas Turbine Instrumentation Conference,* September, 2009, Norrköping, Sweden

Grzybowski, R. (1998). *Microwave recess distance and air-path clearance sensor.* US Patent 5,818,242

Hayes, T.; Hayes, B.; Bynum K. (2011). Utilizing Non-Contact Stress Measurement System (NSMS) as a Health Monitor. *Proceedings of 57th International Instrumentation Symposium,* June 2011, St. Louis, Mo USA

Heath, S. (2000). A New Technique for Identifying Synchronous Resonances Using Tip-Timing. *ASME Journal of Engineering for Gas Turbines and Power.* Vol. 122, No. 2

Holst, T.A. (2005). *Analysis of Spatial Filtering In Phase-Based Microwave Measurements of Turbine Blade Tips.* Georgia Institute of Technology Master Thesis. Atlanta USA

Loftus, P. (2010). *Determination of blade vibration frequencies and/or amplitudes.* US Patent application no. 20100179775

Madden, R. (2010). UK Blade Tip Timing. A customer perspective. *Proceedings of ISA International Instrumentation Symposium,* May 2010, Rochester, NY USA

Millar, R.C. (2011). Turbo-Machinery Monitoring Measures for Propulsion Safety and Affordable Readiness. *Proceedings of 57th International Instrumentation Symposium,* June 2011, St. Louis, Mo USA

Nicholas, T. (2006). *High cycle fatigue: a mechanics of materials perspective,* Elsevier, ISBN 978-0080446912, Oxford, UK

Nyfors, E. (2000). Industrial Microwave Sensors – A Review. *Subsurface Sensing Technologies and Applications,* Vol. 1, No. 1

Poznańska, A. (2000). *Lifetime of Turbojet Blades Made up of the EI-867 Alloy with Consideration of Non-Uniform Deflections and Structural Changes.* PhD Thesis (in Polish). Technical University of Rzeszów

Przysowa, R.; Spychała J. (2008). Health Monitoring of Turbomachinery Based on Blade Tip-Timing and Tip-Clearance. RTO-MP-AVT-157-P14. In: *Ensured Military Platform Availability*. Available from: www.rta.nato.int

Roberts J.P. (2007). Comparison of Tip Timing with strain gauges for rotor blade vibration measurement. In: *Proceedings of Lecture Series on Tip Timing and Tip Clearance Problems in Turbomachines*. Von Karman Institute. Belgium

Rokicki, E.; Weryński, P.; Szczepanik, R.; Spychała, J.; Michalak, S.; Perz, M. & Przysowa, R. (2007). *A differential antenna* (in Polish). Patent application no. P381806

Ross, M.M. (2007). The Potentials Of Tip Timing / Tip Clearance Measurements. The Land Based Power Plant Perspective. In: *Tip Timing and Tip Clearance Problems in Turbomachines*. 2007. VKI LS 2007-03, von Karman Institute, Belgium.

Rushard, P. (2010) A Process for the Application of Blade Tip Timing Measurements on a Gas Turbine. *Proceedings of ISA International Instrumentation Symposium*, May 2010, Rochester, NY USA

Shephard, D.; Tait, P. & King, R. (2000). Foreign Object Detection Using Radar. In: *Aerospace Conference Proceedings*, IEEE, Volume: 6, pp. 43-48

Szczepanik, R. & Witoś, M. (1998). A computer diagnostic system for turbojet engines based on the discrete phase method for measurement of blade vibration (in Polish). In: *Output of Aviation Institute*, No. 1 (152), p. 135-149.

Wagner, M. (1998). Novel microwave vibration monitoring system for industrial power generating turbines. *Proceedings of Microwave Symposium Digest*, IEEE MTT-S International, pp.1211-1214 vol.3

Wenger, J. (1997). An MMIC-based microwave sensor for accurate clearance measurements in aircraft engines. *Proceedings of 27th European Microwave Conference and Exhibition*. Vol. 2, pp. 1122-1126

Witoś, M. (1994). *A Non-Contact Technique of Vibration Measurement Applied to Health Monitoring of Compressor Blades in a Turbine Engine* (in Polish), PhD Thesis, Air Force Institute of Technology (ITWL), Warsaw

Witoś, M. (1999). Blade vibration as a symptom for condition monitoring of aircraft engines. In: *Issues of testing and operation of avionic equipment*. Air Force Institute of Technology (ITWL), Warsaw, vol. 4, pp. 297-315

Witoś, M. & Szczepanik, R. (2005). Turbine Engine Health/Maintenance Status Monitoring with Use of Phase-Discrete Method of Blade Vibration Monitoring, *Proceedings of Symposium on "Evaluation, Control and Prevention of High Cycle Fatigue in Gas Turbine Engines for Land, Sea and Air Vehicles*. RTO-MP-AVT-121-02. Granada, October 2005

Witoś, M. (2007). Theoretical Foundations of Tip Timing Measurements. *Proceedings of Lecture Series on Tip Timing and Tip Clearance Problems in Turbomachines*. Von Karman Institute. Belgium

Zablockij, I. E.; Korostiliew, J.A. & Szipow, R.A. (1977). *Non-contact blade vibration measurement in turbomachinery* (in Russian). Maszinostrojenije. Moscow

Zielinski, M. & Ziller, G. (2005). Noncontact Blade Vibration Measurement System for Aero Engine Application. *Proceedings of 17th International Symposium on Airbreathing Engines*, September, 2005. Munich, Germany, Paper No. ISABE-2005-1220

Żółtowski, B. (1984). Design of experiments in machinery diagnostics (in Polish), In: *Scentific dissertations*, WSO Toruń, Vol. No. 1, pp. 61-64, Toruń

Part 4

Economic and Environmental Aspects

9

An Overview of Financial Aspect
for Thermal Power Plants

Soner Gokten
Gazi University
Turkey

1. Introduction

Thermal Power Plants are facilities that produced electrical energy which is a secondary energy source by using the primary energy sources. In reference to International Energy Outlook Report, marketable energy consumption in the World, taking 2007 as the base year, is expected to grow 49 percent until 2035 (IEO, 2010). According to this report, the distribution of estimated energy consumption by primary sources is shown in Fig. 1.

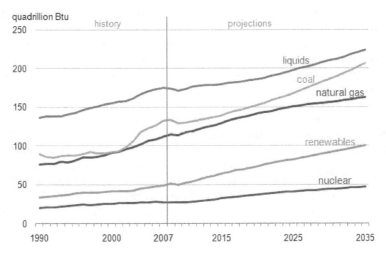

Fig. 1. World marketed energy use by fuel type, (IEO, 2010)

The revealed projection provides a picture concerning what kind of primary sources of energy should be appealed on producing electricity. It is put forward that the production of electricity with an increase of 87 percent until the year 2035 will respectively reach up to 25 trillion in 2020 and to 35.2 trillion kilowatt hours in 2035, taking 2007 as the base year in the report. As can be seen in Figure 2, coal is primary energy source, the most preferred, in production of electricity and natural gas, renewable energy sources and liquid energy sources, respectively, follow it. At present, 80 percent of the World's production of electricity is carried out by fossil fuel power plants (coal, petroleum products, natural gas),

20 percent of the World use the different types of primary energy source like hydraulic, nuclear, wind, solar, geothermal and biogas (SRWE, 2007).

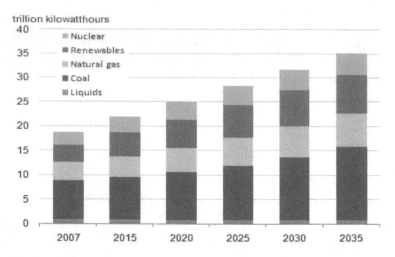

Fig. 2. World net electricity generation by fuel, (IEO, 2010)

Undoubtedly, the choice of the primary sources for electricity production depends on many factors and these factors are fundamental to determine the types of thermal power plants. Thus, these factors also drivers of the decision making processes of financial management function in thermal power plants. Also, the close relation, between financial development and economic growth, increases the importance of financial perspective in management of thermal power plants.

Financial management decision making process is applied on the basis of present value maximization within the framework of the investment and operating costs. Therefore, the factors, influencing decisions related to thermal power plants, affect investments and operating costs indirectly. In this study, it is tried to examine the components that related with financial decision and to present the internal relations between the financial management and thermal power plant in a panoramic view. In the following sections, financial decision making process in thermal power plants are discussed, under the titles of 'The Factors Affecting Decision Making Processes', 'Investment, Operating and Supporting Costs' and 'Long-Term Supply Contracts' respectively. Concluding remarks are then offered.

2. The factors affecting decision-making processes

Energy sector is a dynamic market that contains many guiding factors. Human beings, especially after industrialization, understood the vital importance of the sector and became to mobilize its ability to obtain energy. Playing such a critical role of energy sector made inseparable part of public sector for the developed and developing countries. Namely, energy sector must not be evaluated on the basis of market economics. This kind of effort will not reflect the truth. Because the invisible hand of public is always on the sector, while the private sector rules seem working.

Thermal power plants is one of the most important element of the energy sector and they are masterworks that enable producing electrical energy which can be thought as one of the basic needs of life after water and food. Preference of the thermal power plant's type in electricity production is a big dilemma and prior discussion subject for related parties in recent years. For instance, environmentalists act against fossil-fuelled thermal power plants or nuclear power plants and they try to warn decision-makers about environmental pollution, global warming, carbon emissions etc. There is no doubt that eliminating the existence of such kind of industrial elements, playing a major role in environmental issues seems true, but only with a view of environmentalism because when many other factors were considered, thoughts of environmentalists cannot be accepted in short- or medium-term. The thoughts, agreed and supported by everyone, can be ignored suddenly and quickly because the needs, called energy, especially in economic frame, have a vital importance. Financially; a bird in the hand is better than two in the bush.

All decisions on the type, innovation and improvement efforts of thermal power plants depend on many factors. It is impossible to execute decision making process as taking only one of these factors. The effective factors in the establishment and operation decision of thermal power plants can be classified as follows;

- Government Policies and Preferences
- Environmental Factors
- Macroeconomic Factors
- Research & Development Opportunities

2.1 Government policies and preferences

Countries compete with each other within the framework of the energy production nearly. Energy, especially for the developed countries, is one of the most important factors within the frame of competitive advantage. The energy consumed by the countries is illustrated in the figure below.

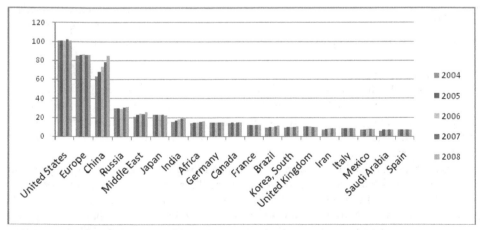

Fig. 3. Countries by energy consumption (Quadrillion Btu), (EIA)

The close relations between industry and energy consumption, especially when looking the data of China, can be observed easily. Namely, energy consumption of Republic of China increased every year from 2004 to 2008. China industry, increasing production capacity, is reportedly building an average of about one coal-fired thermal power plant a week to meet the requirement of its high-level of electrical energy consumption (Figure 4) (Bradsher and Barboza, 2006).

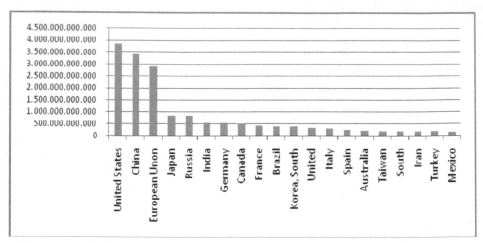

Fig. 4. Countries by electric energy consumption (kwh), (EIA)

Primary energy source possibilities of countries are one of the basic factors that determine the preferences of a thermal power plants. Namely, U.S.A., Germany, India and China produce more than 50% of their electrical energy by coal-fired thermal power plants, while most of the thermal power plants, in the countries has an abundance of natural gas such as Qatar, are gas-fired. This choices are directly related with the reserve capabilities of the primary energy sources which are one of the main issues for government policies and preferences. For example, the coal-fired thermal power plants are preferred to operate reserves, without taking into consideration of performance criteria and environmental factors, although coals has a poor heat value and high carbon content in Turkey.

2.2 Environmental factors
The basic element with energy production and environment interaction is the danger of climate change. Carbon dioxide, methane, nitrous oxide etc. is piling up on the layers of atmosphere, continuing of this situation by increasing prevents the reflection back and faced to a heating risk of the earth. Melting of glaciers and rising sea water are the natural results of warming phenomenon.

Undoubtedly, to tackle with the kind of this problem requires a global consciousness and joint actions. In this direction, the studies begin at the end of the 1980s, were brought to the agenda of the world in 1992 with Rio summit and a joint consciousness was formed by courtesy of the United Nations Framework Convention on Climate Change, accepted on 21st March, 1994. The UNFCCC, embraced a good faith at the point of decreasing its greenhouse gas oscillation based on 1990 by the parties in the contract. However, it seems insufficient because of the poor sanctions and absence of quantitative targets. Kyoto Protocol, accepted in 1997, 3rd Parties Conference of The Agreement of United Nations

Framework Convention on Climate Change and came into operation in 16th February, 2005 is the first concrete step to achieve the ultimate purpose of the contract. Since the protocol includes concrete quantitative targets for the period of 2008-2012 and open sanctions, it creates possibilities for other international agreements to conflict with climate change.

The major role of energy sector in greenhouse gas oscillation as a principal cause of climate change; the measures, to be taken and apply against to such danger, is necessitated predominantly in energy sector. At this point, the most serious press is dragooned by environmental protectionist about fossil fuelled thermal power plants. According to them, fossil fuels should use just a bit in indispensable situation or new substances must be exist for taking substitute. There is no doubt that coal-fired thermal power has the highest rate of carbon emissions among the fossil fuelled thermal power plants. Many countries which want to use their primary energy sources in a most efficient way; invest a lot of money for technological research to develop clean coal, line to keeping carbon emissions, arrested, confinement, isolation, treatment, swallowed, destruction and storage. Several alternative methods investigated continuously (Prisyazhniuk, 2008). Many important leaders reckon on clean coals technologies in order to continue coal-fired based electricity production.

2.3 Macroeconomic factors

The close relationship between economic development level of countries and energy or electric energy consumption is one of important elements which directly effects necessity of thermal power plant. There are a lot of works in literature that try to analyze the relationship between energy or electric energy consumption and economic growth.

The relationship between economic growth and energy consumption was found as casual for Japan by Erol and Yu (1987), for Turkey, France, Germany, Italy, and Korea by Soytas and Sari (2003), for Canada by Ghali and El-sakka (2004), for Taiwan by Holtedahl and Joutz (2004). Lee and Chang (2008) found unidirectional causal relationship in the long-term between energy consumption and economic growth for 16 Asian Countries between 1971-2002 in their works. Moreover, in studies of Lee et al (2008), it was worked on 22 OECD countries base on 1960-2001 period, and as a result, a bi-directional causal relationship was found between energy consumption, capital, and GDP. Confessed works are very important, especially, for policy makers. Because, way of relationship between energy consumption and economic growth represents various options to decision-makers related to interaction between energy and GDP. Important works related to determination of the relationship between energy consumption and economic growths were represented chronologically in Table 1. These works, at the same time, suggest to be understood what energy dependency is in economically.

Mentioning about energy dependency needs to accept energy consumption as the most important part of economic growth. In other words, we add energy consumption between capital and work elements which are basic inputs of economic growth and accept that a decline in energy consumption effects real GDP negatively. To confirm this kind of hypothesis, causality must be realized as runs from energy consumption to economic growth. On such an occasion, country economy is expressed dependent on energy, and decision-makers try to prevent the negative effect on real GDP to apply energy conservation

Researcher(s)	Method	Countries	Result
Kraft and Kraft (1978)	Bivar. Sims Causality	USA	Growth → Energy
Yu and Choi (1985)	Bivar. Granger test	South Korea	Growth → Energy
		Philippines	Energy → Growth
Erol and Yu (1987)	Bivar. Granger test	USA	Energy ~ Growth
Yu and Jin (1992)	Bivar. Granger test	USA	Energy ~ Growth
Masih and Masih (1996)	Trivar. VECM	Malaysia, Singapore & Philippines	Energy ~ Growth
		India	Energy → Growth
		Indonesia	Growth → Energy
		Pakistan	Energy ↔ Growth
Glasure and Lee (1998)	Bivar. VECM	South Korea & Singapore	Energy ↔ Growth
Masih and Masih (1998)	Trivar. VECM	Sri Lanka & Thailand	Energy → Growth
Asafu-Adjaye (2000)	Trivar. VECM	India & Indonesia	Energy → Growth
		Thailand&Philippines	Energy ↔ Growth
Hondroyiannis et al. (2002)	Trivar. VECM	Greece	Energy ↔ Growth
Soytas and Sari (2003)	Bivar. VECM	Argentina	Energy ↔ Growth
		South Korea	Growth → Energy
		Turkey	Energy → Growth
		Indonesia & Poland	Energy ↔ Growth
		Canada, USA & UK	Energy ↔ Growth
Fatai et al. (2004)	Bivar. Toda and Yamamoto (1995)	Indonesia & India	Energy → Growth
		Thailand&Philippines	Energy ↔ Growth
Oh and Lee (2004b)	Trivar. VECM	South Korea	Energy ↔ Growth
Wolde-Rufael (2004)	Bivar. Toda and Yamamoto (1995)	Shanghai	Energy → Growth
Lee (2005)	Trivar. Panel VECM	18 developing nations	Energy → Growth
Al-Iriani (2006)	Bivar. Panel VECM	Gulf Cooperation C.	Growth → Energy
Lee and Chang (2008)	Mulitv. Panel VECM	16 Asian countries	Energy → Growth
Lee et al. (2008)	Trivar. Panel VECM	22 OECD countries	Energy ↔ Growth
Narayan and Smyth (2008)	Multiv. Panel VECM	G7 countries	Energy → Growth
Apergis and Payne (2009)	Multiv. Panel VECM	11 countries of the Commonwealth of Independent States	Energy ↔ Growth
Apergis and Payne (2009)	Multiv. Panel VECM	6 Central American countries	Energy → Growth
Lee and Lee (2010)	Multiv. Panel VECM	25 OECD countries	Energy ↔ Growth

Table 1. Overview of Selected Studies, (Belke et al., 2010)

policies. On the contrary, if the unidirectional causality runs from real GDP to energy consumption, decrease in the energy consumption will not affect real GDP significantly. If bi-directional causality happens between energy consumption and economic growth, energy consumption and real GDP effects each other simultaneously. This kind of interaction directs policy makers to take measures on energy use. But, at the point of declining energy use in arrangements, it is necessary to separate industrial dimension of energy use to avoid from potential negative effects of energy consumption on real GDP. In this frame, decision-makers can assume that to head for more efficient energy sources or to use energy generating technologies which creates less environmental pollution as an alternative.

The basic relationships, mentioned above, represent a macroeconomic point of view related with which the reasons the countries faced to energy dependency or had bi-directional causality between economic growth and energy consumption, prefer thermal power plant. Because, each country is not able to fulfil the Kyoto Protocol completely and it seems impossible to quit thermal power plants with fossil fuel for some countries in economic frame. Primary sources are the most important inputs for production process in thermal stations. For this reason, fuel costs are important economic factors for coal, nuclear, and natural gas based thermal power plants. The thermal power plants, use solar, geothermal, and wind energy, are excluded because of none of fuel cost. Historical costs related to coal and natural gas costs are represented in the figure 5 and 6 respectively. Uncertainties of costs of primary sources using as a fuel in electric production have critical effects on policies that are applied by the countries in energy sector. The easiest way to manage mentioned uncertainties is to build thermal power plants according to domestically existing primary sources of the countries. Otherwise, management effectiveness cannot be provided in the relationship between energy consumption and economic growth.

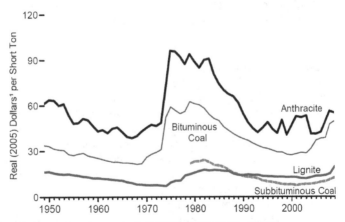

Fig. 5. Coal prices by type for the period of 1949-2009, (AER, 2009)

Countries which have high-level coal reserve still prefer coal-fired thermal power plants in primary level despite with environmental problems. Costs of coals, used as fuel are directly dependent on calorie rate which determines heat capacity of coal. Nonetheless, the inorganic elements within coal caused environmental pollution. Anthracite has highest heat rate and least ash and poisonous exodus for environmental pollution, is a coal. Nonetheless, reserve size of it is much less than the reserve size of lignite because, it's history bases on 300 million years ago. Water and a lot of foreign substance exist within lignite, approximately started to form 60 million years ago. Therefore, it shows a increasing feature of environment pollutions.

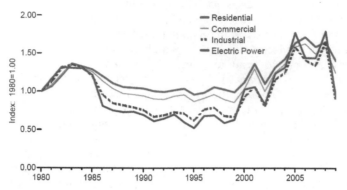

Fig. 6. Natural gas indexed prices by sector for the period of 1967-2009, (AER, 2009)

As it has been seen in the figures, natural gas price which is used as fuel in electric production, is especially more unstable than coal price. Especially, lignite price has almost been stable during years. It will be able to be understood why coal is more preferable in energy production, if it is taken into consideration that coal price is much less than natural gas cost. Leaving coal-fired thermal power plants, because of environmental pollution, for a country which has high-level lignite reserve, is as unreasonable as depending on an uncontrolled price formation while built gas-fired thermal power plant for a country which has not enough natural gas reserve. Dilemma in this example becomes more complex when all factors are handled together about thermal power plant management and its construction.

2.4 Research & development opportunities

The Thermal power plants, besides being the facilities producing electricity, are one of the towering sectors in the world with the technologies they have. Countries positioning their research and development strategies in the thermal power plant construction and formation frame work in a right way derives revenue by selling the know how knowledge they got to other countries. In this framework, countries sometimes pioneer various thermal power plant investments for creating research and development opportunities and bringing the related technologies in their countries or they prompt the related field by making required arrangements related with decisions of thermal power plant type in their countries.

Nuclear Power plant is in the position of an important laboratory especially for the countries wanting to own nuclear technology. Up until now, many countries utilize the nuclear power plants as a device for the studies of developing nuclear technology or they swell the number of them.

Countries that can observe the future in a right way, namely the ones having vision, seriously make investments to research and development activities on types of thermal power plants, moreover they try to switch the energy production in their countries to the stated technology in spite of the potential damage risks. One of the most important sample of this is Spain. In Spain, among the towering leaders in renewable energy field, important studies are made especially in solar energy field. Forasmuch as, Spain is in the position of an important producer and exporter country in photovoltaic industry.

At the root of this success of Spain, there is its displaying the special importance it gave to renewable energy sources with Renewable Energy Plan considering the period between 2005-2010 as a government policy and paving the way of research and development activities by switching solar thermal power plants. In accordance with Renewable Energy Plan big importance is given to developing photovoltaic batteries and also producing energy with this way. According to this plan, supporting R&D activities that develop the evolution, producing, trading and setting periods of photovoltaic battery technologies is decided. Spanish government gives the purchase guaranty of the electricity produced from the renewable energy sources and subsidizes in system installations.

Another field in which research and development studies gain speed is also wind power based thermal power plants. Germany, a world leader in wind energy, belongs to approximately one –third of total production capacity of the world. Australia, Canada, China, France, India, Italy, Philippines, Poland, Turkey, England, and USA can be showed as the other countries that can lead utilizing the wind power in the world.

3. Investment, operating and supporting costs

We tried to examine the factors affecting the decisions on thermal power plants: 'Government Policies and Preferences, Environmental Factors, Macroeconomic factors and Research and Development Opportunities'. Appearance of stated effects on financial side realize via costs.

These costs can be handled into three groups;
- Investment Costs
- Operating Costs
- Supporting Costs

Decision process on need of thermal power plants, type selection, establishment decision, managerial factors, etc. are carried on in financial framework over the three cost group stated above. Each cost factors are affected from the factors, details of which we expressed, and jointly make out a financial bill to decision makers. Figure 7 illustrates that in what kind of interaction the financial point of view in decision process is happened.

3.1 Investment costs

Investment cost covers engineering, procurement, construction, transmission and capitalized financing costs within. Engineering, procurement and construction costs include preparing the project of the stated facility, providing required equipment and materials and all kind of costs related with construction activities following the decision of establishing thermal power plant. Type of thermal power plant is the basic phenomenon determining the capacity of cost item occurring from these three factors. In general, Nuclear power plant is the type of thermal power having the highest investment cost and respectively coal-fired, wind or solar and at last row gas-fired thermal power plants follow this. Moreover, increases in price of construction materials as iron and steel that show raw material and semi finished material features, increasing demands on machines and equipments used in electricity production facilities, increases in occupational wages and lack of educated manpower recently cause continuously increasing in these three costs.

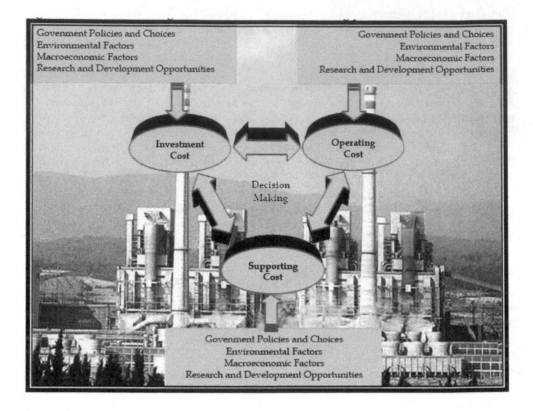

Fig. 7. Interaction diagram of decision making process

Transmission costs are obligatory investments of thermal power plant establisher in primary source supply and distribution of produced electricity framework. How to make the transmission especially in primary source supply framework must be meticulously appraised in feasibility studies. Namely, basic reason that the coal-fired thermal power plants are established close to coal mines, nuclear power plants are established in seaside and lakesides by reason of water need etc. is decreasing required transmission investment costs.

Financing costs endured for supply of required fund for establishing thermal power plant are included in investment cost by being capitalized in framework of feasibility studies. Interest and equity costs take place in it.

3.2 Operating costs

Operating costs consist of expenditures that thermal power plant endures in order to fulfill the main activity. Stated costs are grouped in themselves as direct labor cost, direct material cost, manufacturing overhead cost, marketing cost, and administrative cost. Direct labour cost includes the wages directly paid to working staff, direct material cost includes the expenditures for raw and semi-finished materials, manufacturing overhead cost includes direct or indirect expenses on production. These three cost groups make the producer reach the unit production cost. In addition to this, as it can be understood from their names, marketing cost and administrative cost include the expenses on marketing and administrative functions. All costs can show fixed, variable or semi-variable features.

The most important cost item in operating costs of thermal power plants framework is the primary source of energy, in other words fuel expenses. Fuel expenses are in direct material costs. While the ratio of stated cost for thermal power plants using natural gas and fuel oil in the total operating cost is around ninety percent, this ratio is remained around at the level sixty per cent to seventies for coal-fired thermal power plants. There is no fuel cost in the thermal power plants using renewable energy such as hydraulic, solar and wind energy.

Fuel cost means charging the production cost of the primary energy source cumulatively into the production cost of thermal power plants. Namely, for example, the coal price of coal-fired thermal power plants using coal mine also includes expenses endured for removal of the coal from the mine. For this reason, thermal power plants highly depend on reserve status, manufacturing activity, price formations of the primary energy source.

3.3 Supporting costs

Among the factors affecting decisions on thermal power plants, costs formed depending on environmental, strategic, socio-political and socio-economic all kind of factors, not for the influence of affecting investment and operating costs, must be evaluated in this group.

Two above-mentioned groups are often introduced in thermal power plant decisions and supporting costs are ignored. Many cost-create events and elements as activities of environmentalists, revision of energy policy in view of natural disasters, building new facilities in order to induce to use clean coal because of environmental factors, supporting the R&D activities needed for developing an energy production technology that will be standing out in the future can be evaluated in supporting cost group. Two basic features of supporting costs can be mentioned: (1) These costs are generally invisible or less visible. In other words, it is difficult to take into an account, because it shows facility that can be calculated depending on several possible scenarios. Therefore, the effectiveness of the energy sector could be increased if they could be planned before they occur. For example; when it is started to establish coal-fired thermal power plants, in what extend is it thought to meet with Kyoto Protocol? (2) These costs can show a negative or a positive feature for the countries. In other words, it can turn into sunk cost according to the way of occurring the event and it can create yields. Spain's research & development costs spent for departure of solar thermal power plants, has not been sunk costs and started to turn into yields within the framework of exports. For this reason, all possible event that can occur while decisions of thermal power plants are being financially evaluated should be analyzed.

4. Long-term supply contracts

The construction decisions of thermal power are taken in the frame of strategic plans and long-term supply contracts made between primary energy source producers and thermal power plant operators. The main reason for this is that each one of them wants to minimize risks depending on fluctuations in the spot market (Joskow,1987,1988,1990; Williamson, 2000).

If the members of long-term supply contracts are both state-owned companies, occurred disagreements between these companies can be settled according to directives of the authority. In this sense the contract price cannot reflect world prices. However, potential private investors seek a contract with powerful sanctions and governance mechanism for the exchange (Uner et al., 2008).

Here, we will try to search the answer of the question ' What kind of a long-term supply contract should be made between thermal power plant and the primary energy source, in condition of the fuel supplier?' Each of following sub-sections briefly describes the parts are to be included in long-term supply contracts.

4.1 Maturity of contract and suspensory conditions
Maturity of supply contracts can be formed as 5 years, 10 years and multiplies of them. However, because the supply contracts cover long periods, many substances taken place in the contract show revisable qualifications. The basic element is that the number of factors which may be needed to be revised will also increase in parallel with prolongation of the maturity of the contract. Because the longer-term increases uncertainties about the future, which increases the number of condition and requirement in making the value chain reached at an optimum level the framework of increasing the number of terms and conditions.

Depending on prolongation of supply contract, transforming many factors into revisable form is not an desired condition. The main reason for this is the paradox of making political risks increased, despite the minimization of operational risks increased. Also, increase in revisable factors will increase the numbers of correction and/or adaptation transactions. In the long term, each correction and adaptation work can have possible negotiation request its train. Negotiations are source of political risks and they prevent the job from being done in a systematically continuous way. For this reason, determination of maturity for long-term supply contracts takes place at the beginning of the most important decision items. It is desired in maturity decision that contracts which do not create political risks in parties, do not have any revisable factors more than required and take into consideration the possible stability period on market in which the transactions will be realized.

"First delivery date" taking place in the parts on long term supply contracts is accepted as maturity beginning of the contract. So, signing date expresses the starting point of the "first delivery date" in association of activities with maturity framework while giving effect to the contract in legal framework. Occations in which induring of contracts clauses are subjected are described as suspensory conditions. Betrayed obligation belonged to parties within the context of contracts should be fulfilled in the whole suspensory conditions as of the singing date. In other words, parties, must provide the conditions presented as suspensory conditions ,during the contract. Suspensory conditions determined for seller exposes the qualification criteria of the seller for works and products.

Suspensory conditions determined for purchaser are formed especially according to the using aim of primary energy source mentioned in the contract. Such that, at the same time, for what the fuel is bought by purchaser, in which test and how it will be used and what kind of production process will it be met form the criteria that the purchaser must have. The importance of the suspensory conditions that will exposed for purchaser is that supply contract is signable before start of the work. Thus, that the work for which fuel will be used as raw material have high operating leverage makes necessary to produce solution for supply trouble before facility construction of the purchaser .Hence, purchaser is responsible for the facility that he will establish because he is to determine "the first delivery date " following the signing of the contract.

4.2 Sale and purchase obligations

Determination of moral rules of commertial processes in long term supply contracts is necessary to make supply relationship effective in long term. While responsibilities evaluated in Sale and Purchase Obligations concepts framework regulate alternative supply resources, using fields of the subject product of commertial processing and similar subject for purchaser; for seller, they regulate the points as stock control, supply mechanism.

Minimization of the operational risk possessed by continual production activity presents importance in long term supply contracts. According to this, enterprise which will use its fuel as a raw material will need continuousness of input to provide continuity in production. For this reason, the purchaser's amount of input stock must be follawable for seller, deliveries and supply contact network must be processed according to purchaser's stock. Long term supply contracts must guarantee the fuel need of purchaser. Especially, if fuel supply cannot be realized due to the negativenesses occurring except force majeure, providing the short term need of purchaser is needed. Stated necessity makes defraying the loss and similar arrangements obligatory.

4.3 Amount and delivery of fuel

Amount of related fuel is called contract amount in long term supply contracts. Naturally, contract amount forms the important part of supply contracts.

Contract amount expresses total amount of fuel supplied during a whole working year. Calendar of working year can be determined either standard 12 months or less than 12 months. Hence, the amount of contract amount per a month is accepted as " planned amount-PM". So, contract amount is the main factor for determination of delivered fuel amount. Such that, contract amount is stable factor especially in planning the monthly deserving. In this context, in order to protect the parties' obligations, contract amount and planned amount do not change. In other words, amount of fuel supplied during the working year cannot be less than contract amount.

4.4 Fuel quality

Clearly expressing the quality of fuel introduced in long term supply contract as well as maturity and pricing one of the most important factors of the contract.

Defining the quality factor as gaps instead of a clear definition make the effective practice of the contract possible. For example, three gaps can be determined for coal according to calorie value. From higher to lower, these can be described as (1) Incitement, (2)Normal, (3)Discount. Price is not revised in normal gap. It is increased in incitement gap (Premium);

it is decreased in discount gap (fine). Thanks to this, an mutually acceptable price fixing for purchaser and seller is created.

4.5 Fuel price

The most importing issue in the long-term supply contracts is pricing. Pricing can be done by using three approach, called: cost approach, market approach and income approach.

The necessity of certainly reflecting the production costs of the enterprise that produce the primary energy source to the price is the thing wanted to express by cost approach. It is reached to this price as a result of profit margin that will be included at the base price by using of income approach. Base price shows a price feature that can be fixed according to conjectural movements in the market and in this framework, it forms the beginning period price of the contract. Besides, it is corrected in the periods determined in the contract and correction is realized in market approach framework. At the base of the market approach, prices of all fuels that can be equal to the stated fuel and all other macroeconomic factors can be evaluated in correction of base price.

4.6 Payments and collections

It is the part of long term supply contracts that organizes the decisions practiced by both parties as purchaser and seller. Delivering of the fuel in delivery point and doing the quality measurements make data flow obligatory together. So, preparing the bills show that the parties fulfil the obligations mutually in the consequence of the processes is highly important to provide this data flow.

Generally, Purchaser is obliged to advance money to the seller in exchange for specific ratio of the fuel and /or show bank guaranty. By way of requital, seller is obliged to deliver the fuel promised to the purchaser and document the process. After fulfilling the mutual obligations and finishing the contract, seller are to give the advance taken from purchaser and bank guaranty back.

4.7 Force majeure

The most important feature of long term contracts is that each clause is adjudicated in usual, considerable conditions. Howsoever, realization of the activities and contracts planned in long term has risks within, initially political risk, as is due the structure of long term contracts, these risks are wanted to be lowered to minimal level.

Force majeure concept is one of the main concepts of law and practice of it is seen in all branches of law. Force majeure is an extraordinary, incidental event, fact, condition that blocks the fulfilment of a responsibility partly or entirely, permanently or temporarily, due to this feature cancelling or delaying the responsibility its fulfilment and maturity or changing the feature of the responsibility, unexpected and unpredictable; even if expected or predicted, that cannot be blocked. Even if in long term supply contracts precautions are taken for the risks that are possible to come across depending on ordinary conditions, it is possible to come across with extraordinary conditions.

5. Conclusion

Thermal power plants are the masterpieces realizing the production of electricity, sine qua non need of our world. The common aim embraced by everyone is preferring thermal power plants realizing the clearest production. However, various factors sometimes militate and they conduct the decision making process on thermal power plants contrary to the environmental sensations. It is needed to look the picture with a financial point of view to understand the reasons of this.

Nearly all of the decisions on type and administration of thermal power plants are made on gathering at the cost base in other words; at finance base. Therefore, it is needed to manage the factors affecting decision-making process for clearer electricity production, transform the supportive costs into investments that can create yields.

6. Acknowledgment

I would like to express my gratitude to Aslihan Akin and Eray Karaarslan who are from University of Turkish Aeronautical Association for their assistance.

7. References

AER, (2009). Annual Energy Review 2009. *Energy Information Administration*, June 2009.

Belke, A.H., Dreger, C. and Frauke, H. (2010). Energy Consumption and Economic Growth - New Insights into the Cointegration Relationship. *Ruhr Economic Paper No. 190*; *DIW Berlin Discussion Paper No. 1017*, June 1 2010.

Bradsher, K. and Barboza D., (2006). Pollution From Chinese Coal Casts a Global Shadow. *The New York Times*, June 11, 2006.

Erol,U. and Yu, E.S.H., (1987). On the causal relationship between energy and income for industrializing countries. *The Journal of Energy and Development*, 1987, 13:113-123.

Ghali, K.H. and El-sakka, M.I.T., (2004). Energy use and output growth in Canada: A multivariate co-integration analysis. *Energy Economics*, 26 (2004) 225–238.

Holtedahl, P. and Joutz F.L., (2004). Residential electricity demand in Taiwan. *Energy Economics*, 26 (2004) 201–224.

IEO, (2010). International Energy Outlook 2010. *Energy Information Administration*, July 2010, DOE/EIA-0484(2010).

Joskow, P.L., (1987). Contract duration and relationship specific investments. *American Economic Review*, 77 (1), 168–185.

Joskow, P.L., (1988). Price adjustment in long term contracts: the case of coal. *Journal of Law and Economics*, 31(1), 47–83.

Joskow, P.L., (1990). Price adjustment in long term contracts: further evidence from coal markets. *RAND Journal of Economics*, 21 (2),251–274.

Lee, C., C. Chang, and P. Chen, (2008). Energy-income causality in OECD countries revisited: The key role of capital stock. *Energy Economics* 30(5), 2359–2373.

Lee, C.C. and Chang, C.P., (2008). Energy consumption and economic growth in Asian economies: A more comprehensive analysis using panel data. *Resource and Energy Economics*, 30, 50- 65, 2008.

Prisyazhniuk, V.A., (2008). Alternative trends in development of thermal power plants. *Applied Thermal Engineering* 28 (2008) 190–194.

Soytas, U. and Sari, R., (2003). Energy consumption and GDP: causality relationship in G-7 countries and emerging markets. *Energy Economics*, 25, 33-37, 2003.

SRWE, (2007). Statistical Review of World Energy. *The British Petroleum Company*, 2007.

Uner, M. M., Kose, N., Gokten, S., and P. Okan, (2008). Financial and Economic Factors Affecting the Lignite Prices in Turkey: An Analysis of Soma and Can Lignite. *Resources Policy*, Vol. 33, No. 4, 2008, pp. 230-239.

Williamson, O., (2000). The new institutional economics: taking stock, looking ahead. *Journal of Economic Literature*, 38 (3), 595–613.

10

A Review on Technologies for Reducing CO_2 Emission from Coal Fired Power Plants

S. Moazzem, M.G. Rasul and M.M.K. Khan

School of Engineering and Built Environment, Faculty of Sciences, Engineering and Health, Central Queensland University, Rockhampton, Queensland Australia

1. Introduction

In recent years, global warming has been a major issue due to continuous growth of greenhouse gas emissions from different sources. It has been estimated that the global average temperature will rise between 1.4 –5.8 °C by the year 2100 (Williams, 2002). The contributors to greenhouse effects are carbon dioxide (CO_2), chlorofluorocarbons (CFCs), methane (CH_4), and nitrous oxide (N_2O). The contribution of each gas to the greenhouse effects is CO_2- 55%, CFCs - 24%, CH_4 - 15%, and N_2O - 6% (Demirbas, 2008). Carbon dioxide (CO_2), a major greenhouse gas which is mainly blamed for global warming occupies a large volume of the total emissions. Figure 1 shows the trend of CO_2 emissions over the years (Demirbas, 2005).

Different industrial processes such as power plants, oil refineries, fertiliser, cement and steel plants are the main contributors of CO_2 emissions. Fossil fuels such as coal, oil and natural gas are the main energy sources of power generation and will continue to generate power due to the large reserves and affordability. It is expected that coal utilisation in power generation will continue to increase in this century too. Demirbas (2005) reported that about 98% of CO_2 emissions result from fossil fuel combustion, and 30%–40% of world CO_2 emissions are generated by coal combustion among all the fossil fuels.

The coal fired power plants generate the majority of the electricity and produce the highest rate of CO_2 per kilowatt hour (Department of Energy and Environmental Protection Agency, Washington DC, 2000). Table 1 shows the CO_2 emissions and power generation from various sources. It can be seen from Table 1 that the coal-fired power plants are responsible for a large percentage of CO_2 emission among other process plants around the world, including Australia. About 46% of the world's power generation is estimated to be from coal combustion, including 50%, 89% and 81% of the electricity generated in the United States, China and India respectively (Parker *et. al.*, 2008). It is estimated that combustion of coal for power generation will be responsible for about 41% of the world's CO_2 emissions by 2025. Table 2 shows the world wide large stationary sources of CO_2 emissions and Figure 2 shows the stationary sources of CO_2 in Australia.

In Australia 75% of the total electricity is produced from coal fired power plants, so coal fired power plants have a great impact on the Australian economy (ESAA, 2003; ABS, 2001). In Australia, power plants contribute about 64% of the total CO_2 emissions generated from stationary sources (AGO, 2006).

Technology	CO_2 Emissions (Kg/MWh)
Pulverised Coal-fired subcritical	850
PC-fired supercritical	800
IGCC	670
NGCC	370
Nuclear	0

Table 1. CO_2 Emissions from Various Power Generation Technologies (Narula *et al.*, 2002)

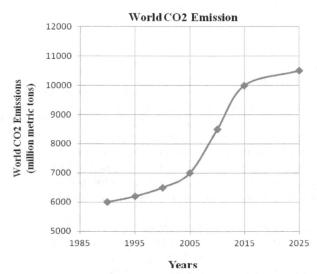

Fig. 1. World CO_2 emissions between 1990 and 2020 (Demirbas, 2005)

Process	Number of sources	Emissions ($MtCO_2$ yr $^{-2}$)
Fossil Fuels		
Power	4,942	10,539
Cement production	1,175	932
Refineries	638	798
Iron and steel industry	269	646
Petrochemical industry	470	379
Oil and gas processing	N/A	50
Other sources	90	33
Biomass Bioethical and bio energy	3,03	91
Total	7,887	13,466

Table 2. The world wide large stationary sources of CO_2 with emissions (IPCC, 2005)

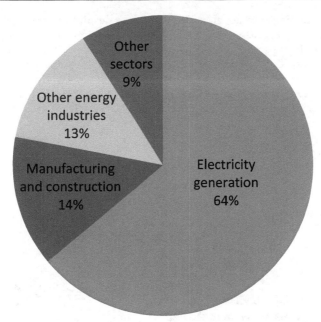

Fig. 2. Stationary sources of CO_2 in Australia (AGO, 2006)

Continued use of coal to produce electricity makes it very important to undertake a study on coal fired power plants with the aim of reducing the hazardous emissions of CO_2 as well as sulphur oxides (SO_x), nitrogen oxides (NO_x) and other particulates to help maintain a sustainable environment. As such, it is of great importance to reduce CO_2 in the atmosphere by reducing emissions from power plants.

There are many technologies available to reduce the emissions of sulphur oxides (SO_x), nitrogen oxides (NO_x) and other particulates, but very little consideration has been given to the reduction of CO_2 emissions (Coal Industry Advisory Board, 1994). Therefore further study on how to reduce CO_2 emissions from a coal fired power plant is currently an important field of research. Clearly this study on different existing CO_2 emission reduction technologies is needed to identify some measures that could be used for the successful reduction of CO_2 emissions from a typical coal-fired power plant. This chapter aims to identify an appropriate technology of CO_2 emission reduction to maintain sustainable environment.

2. Introduction to coal fired power plant

Electricity can be produced by various sources such as fossil fuels, nuclear fission, renewable sources etc. Figure 3 shows a typical schematic diagram of electricity production from burning coal. A typical pulverized coal (PC) combustion power plant is equipped with three units, boiler block, generator block and flue gas clean up block. The boiler block is the main unit where coal is burned with air to generate high pressure steam; the generator block contains the steam turbine/electric generator set, condenser and cooling water; and the third block is the flue gas clean-up unit which removes particulate matter (PM) and other pollutants from the flue gas to control emissions. This third unit carries out Selective Catalytic Reduction (SCR) for NOx (Nitrogen Oxide) removal, electrostatic precipitation

(ESP) for particulate matter removal, and wet flue gas desulphurisation (FGD) or wet lime scrubbing to remove SOx and mercury. The level of emission control of this unit is 95% – 99% depending on the type of coal used (World Coal Institute, 2010). Narula illustrated that, due to the addition of CO_2 amine scrubbers at the back end of the power plant to reduce CO_2 emissions, the net plant output decreases by about 25 percent from 2 x 400 to 600 MW, and the plant heat rate increases to 13,250 kJ/kWh from 9,800 kJ/kWh, and the capital cost of the plant increases by about 77% (Narula et al., 2002).

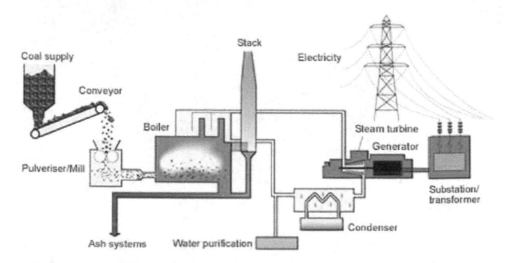

Fig. 3. Diagram of electricity generation from coal (World Coal Institute, 2010)

3. Carbon Capture and Storage (CCS) technologies

To prevent major climate change, CO_2 concentration in the atmosphere should be reduced by either CO_2 up-take from the atmosphere biologically or reducing the CO_2 emissions from the sources. There are some approaches available for reduction of CO_2 emissions from stationary sources such as reduction of the consumption of energy generated using fossil fuels, increase in energy generation by non-fossil fuel sources such as solar, wind, biomass, and nuclear energy and using carbon capture and storage (CCS) technology for large scale production. In CCS-technologies, CO_2 is separated from the flue gas from any source and used in other processes or stored in a safe place, such as underground storage and ocean storage. In this study, only reducing/capturing CO_2 from flue gas will be considered among the three phases (capture, transport, storage) of CCS technology.

The idea of separating and storing CO_2 for mitigation of its emissions to the atmosphere was first proposed in 1977 (Marchetti, 1977). Since then a lot of research work has been done on the possible mitigation options. Nowadays there are many CO_2 capture technologies available; some of these technologies are commercially established and some are under development. Mainly there are three pathways (illustrated in Figure 4) to reduce CO_2 emissions, these being post or after combustion, pre or before combustion and oxy-fuel combustion with CO_2. In pre-combustion processes, CO_2 and other pollutants such as NOx and SOx are removed through gasification before combustion (Kreutz et al., 2002; Williams,

2003). On the other hand, CO_2 is removed after combustion in post combustion technology. In oxy-fuel combustion, CO_2 is separated during combustion generating a flue gas stream containing mainly CO_2 and H_2O. This technique is simple and comprises mainly compression and cooling steps and no extra solvents are required (IPCC, 2005). This technology is mainly used in glass, aluminium and steel furnaces to remove CO_2, but in power generation it is still an emerging technology and some large scale pilot plants are planned or under way.

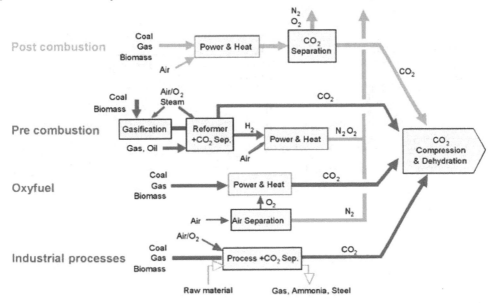

Fig. 4. CO_2 capture technologies (IPCC, 2005)

3.1 Pre-combustion CO₂ capture

This process removes the CO_2 from any industrial sources prior to combustion of fuel like coal, oil or gas to produce energy. In the pre-combustion process, fuel is first converted into synthesis gas containing hydrogen and carbon monoxide (CO). This CO reacts with water and produces CO_2, and finally this CO_2 is separated from the hydrogen and compressed for transportation and storage. Then the remaining hydrogen is combusted to produce energy. About 90%-95% of CO_2 emissions can be reduced by this technology. Pre-combustion technologies are shown in Figures 5. This technology is currently used in oil refineries, but has limited use in power plants. Integrated Gasification Combined Cycle (IGCC) and Fluidized Bed Combustion (FBC) technology are involved in pre-combustion CO_2 capture. Currently, Integrated Gasification Combined-Cycle (IGCC) technology is used to produce electricity and reduce emissions from power plants.

Carbon is captured using IGCC technology before combustion using low pressure with a physical solvent (e.g., Selexol and Rectisol processes), or a chemical solvent (e.g., methyl diethanolaimine (MDEA)). In this process, fossil fuel is first converted into CO_2 and Hydrogen gas (H_2). Then, the H_2 and the CO_2 gas are separated from each other and electricity is produced by the combustion of Hydrogen-rich gas. About 90% of the CO_2 can be removed from a power plant by pre-combustion CO_2 capture using IGCC technology, though pre-combustion technology is mainly applicable for new power plants, not being economic for existing plants. It was found that currently four commercial IGCC plants (each plant has capacity of 250 MW) are operated worldwide. Power plant efficiency is reduced from 38.4% (without CO_2 capture) to 31.2% (with CO_2 capture) by introducing IGCC technology to reduce emissions (MIT study, Future of coal, 2007).

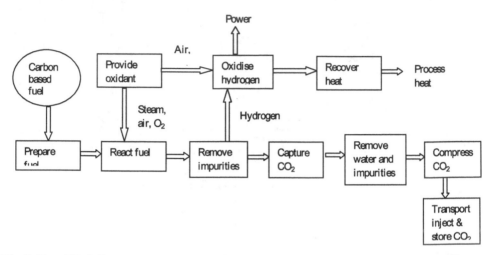

Fig. 5. Simplified illustration (redrawn) of Pre-Combustion CO_2 Capture (Scottish carbon capture and storage, 2010)

3.2 During combustion CO_2 capture or oxy-fuel combustion technology

During combustion CO_2 capture involves combustion of the coal with oxygen (nearly pure oxygen >95%) instead of air. A flue gas is produced consisting mainly of highly concentrated CO_2 and water vapour. These two components of flue gas are easily separated through a cooling process. The water is then condensed and a CO_2 rich gas-stream is formed. This oxy-fuel process can remove up to 100 % CO_2 from the flue gas. The oxy-fuel process is illustrated in Figures 6. The main problem of this technology is that separating oxygen from the air causes energy penalty to the power plant. CCS consumes significant amount of energy. This additional energy is supplied from the power generation cycle causing less amount of energy (electricity) output available or demanding additional amount of energy (as input) to generate same amount of energy (electricity) output. This (energy consumed by CCS) is termed as energy penalty and it ultimately raises the cost of power generation. Chemical looping combustion technology which is under development can potentially remove this problem by more easily separating oxygen from the air.

The Vattenfall Project (30MW pilot plant) in Germany and the Callide Oxyfuel Project in Queensland, Australia are the largest oxy-fuel demonstration projects under development. It was estimated by MIT that, after installation of oxy-fuel technology, power plant efficiency will be reduced by 23% for new construction and 31%-40% for retrofitting in an existing plant (MIT study, Future of coal, 2007).

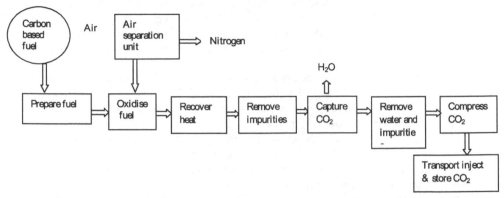

Fig. 6. Simplified illustration (redrawn) of Oxy-Fuel CO₂ Capture (Scottish carbon capture and storage, 2010)

3.3 Post-combustion CO₂ capture

Post-combustion CO₂ capture for power plants takes place after combustion of air and fuel to generate electricity, and immediately before the resulting exhaust gas enters the stack. The advantage of post-combustion CO₂ capture technology is that it can be retrofitted to existing plants without major modifications; only the necessary capturing equipment is required to be installed. A simplified illustration of post combustion CO₂ capture is shown in Figure 7. Several post combustion CO₂ capture methods are available. The most common post combustion method is chemical absorption with amine solvents. Other post combustion CO₂ capture technologies are membranes, the PSA (pressure swing adsorption) process and mineral carbonation processes. These are described below.

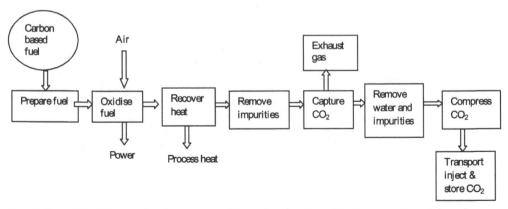

Fig. 7. Simplified illustration (redrawn) of Post Combustion CO₂ Capture (Scottish carbon capture and storage, 2010)

3.3.1 Absorption technology

Up to date the most available and proven capture technology for industrial application is the chemical absorption process using amines (monoethanolamine (MEA)). In this process, exhaust gases containing CO_2 pass through an absorber where CO_2 binds with the MEA solution, and then CO_2-rich MEA is pumped to a stripper for regeneration of the solvent and separation of CO_2 from the MEA. In chemical absorption processes, a chemical bond is formed between gaseous CO_2 and alkaline solvents and, due to this bond formation, chemical absorption processes are kinetically faster. Many commercial electricity generation plants use a chemical absorption process for CO_2 recovery (IEA-GHG, 2000). When CO_2 partial pressure is less than 3.5 bar, then a chemical absorption process is preferred (GPSA, 2004). Figure 8 shows a representation of the chemical absorption process. Flue gas containing CO_2 is cooled down before entering into the absorber which maintains a temperature range of 40-60° C. In the absorber, flue gas contacts with the absorbing solvent and binds with the absorber. Then the lean flue gas leaves the top of the absorber and rich solvent loaded with CO_2 leaves the bottom of the absorber and is pumped to the regenerator to recover the solvent through a heat exchanger. In the regeneration section, rich solvent is heated to release CO_2 from the top of the regenerator and finally CO_2 is compressed for storage. The lean solvent free of CO_2 is reused in the absorber (IPCC, 2005).

A physical absorption process is carried out by the weak binding of CO_2 and the solvent at high pressure. Physical absorption is mainly considered for high CO_2 concentrations (higher than 15%) and high partial pressures. The physical absorption process is still in the preliminary stages of development. Rectisol, Purisol, Selexol, and Fluor solvents are used for physical absorption, and MDEA, KS-1, KS-2, KS-3, MEA, Amine Guard are used in chemical absorption (International Energy Agency, 2004). Less energy is required for solvent regeneration in the physical absorption process compared to chemical absorption.

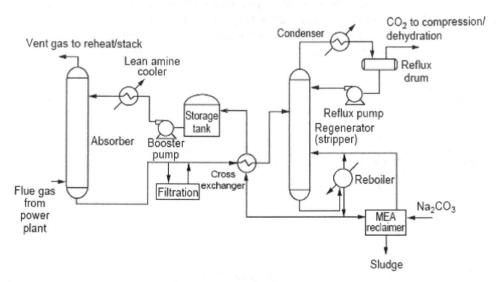

Fig. 8. Amine Chemical Absorption Process (Herzog & Golomb, 2004).

Generally, alkaline solvents such as alkanolamines, hot potassium carbonate, and ammonia are commercially used for CO_2 recovery. Besides these, amine solvents such as monoethanolamine (MEA), diethanolamine (DEA) and methyl diethanolamine (MDEA) are used in chemical absorption processes. But MEA is the most suitable solvent due to having some favourable characteristics such as high solubility in water, high biodegradability , high selectivity, fast reaction kinetics, high affinity of CO$_2$ and being easier to regenerate (DOW, 2005; IPCC, 2005). Hindered-amine solvent KS-1, developed by Mitsubishi Heavy Industries, has lower energy consumption for regeneration (Mimura et al., 1997). Recovery of CO$_2$ is typically 85% to 95% and CO$_2$ purity is approximately 99.5% with amines solvents (IPCC 2005). Recently, it has been estimated by MIT that efficiency losses of 25%-28% for new construction and 36%-42% for retrofitting on an existing plant apply as a result of the installation of MEA CO$_2$ capture technology. In addition, degradation of the amine in the absorber through overheating above 205$\circ$F or through oxidation from various causes are other major drawbacks. Flue gas desulphurisation (FGD) or selective catalytic reduction (SCR) devices are required to be installed with the MEA process for removing these drawbacks (Parker et al., 2008). In the chemical absorption process, solvent choice, solvent cooling, heating, regeneration, absorption, pumping and the compression of the purified CO$_2$ all are energy consuming and costly processes which reduce the overall efficiency of the plant. The absorption column and regeneration unit are both expensive to operate and a high capital investment (Goldthorpe et al., 1992). Therefore, research and developments are needed to improve the energy efficiency, especially in the absorption and regeneration aspects of this technology as well as cooling and heating issues for effective integration of chemical absorption processes into power plants. In addition, choice of solvent is a major factor. The key factors for selection of solvent are heat of absorption/regeneration, CO$_2$ absorption rate, CO$_2$ absorption capacity, resistance to degradation and impurities, corrosion, and volatility (Cullinane et al., 2002). The chemical absorption method using MEA is a very expensive and energy intensive process. Binding between CO$_2$ and solvent molecules is strong and this offers effective removal of most of the CO$_2$. But due to this strong binding, high regeneration energy is required to regenerate the solvent.

A chilled ammonia solvent chemical absorption process has been developed in which the flue gas temperature is reduced from about 54.4 $\circ$C to 1.6-15 $\circ$C before entering the absorber to mitigate oxidation problems. This chilled ammonia process lowers the flue gas temperature which minimises the flue gas volume entering the absorber by condensation of residual water in the gas; this also causes some pollutants in the flue gas to drop out, reducing the need for other upstream control processing, (Parker et al., 2008).

Another ammonia based process designated ECO$_2$ is being developed by Powerspan . In this process in which does not involve chilling, the higher temperature flue gas is used to increase the CO$_2$ absorption rate in the absorber to remove CO$_2$ from flue gas (Powerspan, 2008; Ryan and Donald, 2008). The operation of two commercial demonstration projects of Powerspan's process will start between 2011 and 2012 using a flue gas slipstream equivalent to a 120MW unit from Basin Electric's Antelope Valley Station in North Dakota and a flue gas slipstream equivalent to a 125 MW unit at NRG's W.A. Parish plant in Texas respectively (Parker et al., 2008).

3.3.2 Membrane technology

In membrane process gas absorption membranes are used as contacting devices between a gaseous feed stream and a liquid solvent stream. Figure 9 shows a schematic diagram of the membrane process. In this process a membrane module is placed in a thermo water bath to maintain a constant temperature. The feed gas mixture enters the membrane module when the valve of the mixed gas cylinder is turned on to the desired flow rate, and the mixed-gas stream fed into the fibre lumen of the end of the membrane module at a slightly lower pressure than that of the liquid side to prevent dispersion of gas bubbles into the liquid.

The feed gas mixture is prepared in a gas preparation system to a given concentration based on the partial pressure principle, and the absorbents are prepared in the feed tank with deionizer water to a given concentration. The liquid absorbent is introduced by a gear pump from the solution tank to the shell side of the module. Flow of gas steam and liquid absorbent is measured by flow meter. Then CO_2 of the gas mixture is diffused through the membrane pores into the liquid in the shell side and is absorbed by the absorbent. CO_2 permeation through the membrane depends on the difference in partial pressure of the feed side and the permeate side (liquid side) which is known as the pressure ratio (Ho, 2007). Then the liquid absorbent loaded with CO_2 entered into another solution tank, and the treated gas steam is released from the other end of the module (Lu *et al.*, 2009). Selection of solvent and membrane material is very important. In the membrane process, aqueous solutions of propylene carbonate, diethanolamine (DEA), methyldiethanolamine (MDEA) and piperazine (PZ) are used as an aqueous solution. Mainly polymeric membranes, facilitated transport membranes, molecular sieves membranes and palladium based alloy membranes are used in the separation of CO_2 (Feron, 1992).

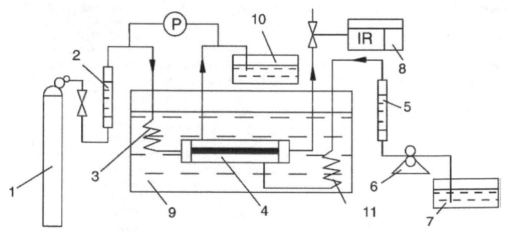

[1 – Mixed-gas cylinder, 2, 5 – flowmeters, 3, 11 – preheater, 4 – membrane contactor, 6 – gear pump, 7, 10 – solution tank, 8 – IR gas analyser, 9 – thermo water bath]

Fig. 9. Schematic diagram of membrane process (Lu *et al.*, 2009)

Aqueous solutions of activated methyldiethanolamine (MDEA) in a PP (polypropylene) hollow fibre membrane contractor showed a better performance in separating CO_2 by the membrane process compared with propylene carbonate with the same membrane contractor (Dindore *et al.*, 2004; Lu *et al.*, 2005). The solvent has low heat of absorption as it requires less energy during regeneration. Another way of improving the chemical absorption process is to ensure that there is maximum interaction between the solvent and the CO_2 (Fei, 2004; Fei and Song, 2005). Other process configurations such as efficient and economic design of absorber, stripper and condenser may be used to improve the process efficiency. Mass transfer coefficient is a very important parameter for the membrane gas process. Mass transfer mainly comprises three steps, diffusion of mixed gas component to the membrane wall, diffusion of the membrane liquid through the pores and finally, dissolution into the liquid absorbent.

Higher operational temperatures and flow rates can enhance the mass transfer coefficient in the absorption membrane process. It also has been observed that a composite amino-acid-based solution (0.75 kmol/m³ GLY salt) + piperazine (PZ) (0.25 kmol/m³) shows better performance than a single amino-acid salt solution, (1.0 kmol/m³ glycine salt, GLY). CO_2 recovery efficiency of this process is approximately 90% (Lu et al., 2009). The membrane process is a highly energy intensive and costly process because of high equipment cost and the high pressure differential required for this process (Herzog et al., 1991). So technology integration is a major issue to improve the membrane materials, their pore size, selectivity and permeability, and a suitable combination of membrane and liquid solvent has the potential for improvement of the economic aspects of this process. An idea of a hybrid configuration with the membrane process has been proposed (Bhide et al., 1998), where a membrane is used with an existing chemical absorption plant at the front side to remove the bulk of the CO_2 and then an aqueous solution of diethanolamine (DEA) is used to remove the remaining CO_2.

3.3.3 PSA (Pressure Swing Adsorption) technology

The diagram of the Pressure Swing Adsorption (PSA) system is shown in Figure 10. The basic principle of the PSA cycle was described by Skarstrom (1960). In the PSA process, high pressure mixed feed gas is fed into the absorber. Then gas with a high affinity to the absorbent is absorbed, and the gas with lower affinity passes through the bed and accumulates at another closed end. When the absorber bed is saturated with high affinity gas, then the outlet valve of the absorber bed is opened and low affinity gas is withdrawn. The pressure of the adsorber bed is reduced as a result, and the absorbed gas is released from the adsorber bed. Finally, the evacuation of the absorbed gas is completed by pumping from the absorber bed. This step may include the purging of the adsorber bed with low affinity gas (Skarstrom, 1960; Ho, 2007)

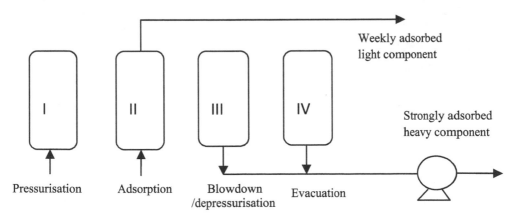

Fig. 10. Schematic diagram of PSA process (Skarstrom, 1960; Ho, 2007)

In Pressure Swing Adsorption (PSA) systems, one of the most important parameters is the choice of the adsorbent for gas adsorption. Adsorbents with high surface areas such as zeolite, molecular sieves and activated carbon have been widely analysed for their CO_2

separation effectiveness. It has been evaluated that zeolite 13X is a better adsorbent for CO_2 as it has higher working capacity, lower purge requirements and higher equilibrium selectivity than activated carbon (Chue et al.,1995). In PSA processes, mainly zeolite 13X is used as the absorbent. Currently it is used in steel and lime industries to reduce CO_2 from flue steams.

3.3.4 Mineral carbonation technology

In mineral carbonation CO_2 comes from different sources react with calcium (Ca) or magnesium (Mg) based natural silicate minerals and form naturally stable solid carbonated product. The storage capacity of this carbonated product is very large compared to other storage options. The carbonated products are stored at an environmentally suitable location or reuse in another industrial process such as mine reclamation and construction, also can be reused and disposal for land filling. It was found from literature that mineral carbonation technology is still in the research and development stage and further research is required to demonstrate and implement this process in power plant (IPCC, 2005). Carbonation of metal oxide bearing minerals with atmospheric CO_2 is a natural process that occurs spontaneously at low partial pressure and at ambient temperature, though this natural process is relatively slows. Based on this natural process subsequent researches are continuing to accelerate the reaction process to introduce this technology in an industrial scale. The idea of CO_2 sequestration by mineral carbonation is relatively new. It was first proposed in 1990 (Seifritz, 1990) and the first published study on this idea was 1995 by Lackner (Lackner et al., 1995). In this process gaseous CO_2 is converted into geologically stable carbonates. Figure 11 shows the schematic diagram of the carbonation process.

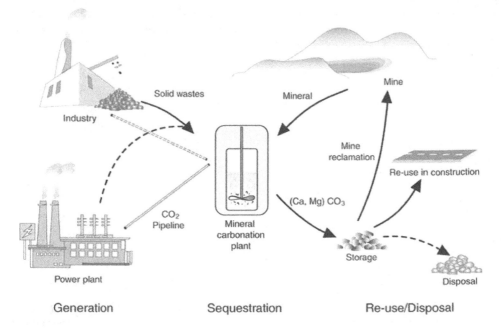

Fig. 11. Schematic drawing of a mineral CO_2 sequestration process (Kojima et al., 1997).

The simple chemical equation of this process is demonstrated below (Lackner *et al.*, 1995),

$$CaO + CO_2 \rightarrow CaCO_3 + 179 \; kJ/mole$$

$$MgO + CO_2 \rightarrow MgCO_3 + 118 \; kJ/mole$$

Figure 12 shows the flow diagram of mineral carbonation process with serpentine and olivine. In this diagram carbonation is carried out after different pre-treatment options such as crushing, grinding, magnetic separation and heat treatment. Size reduction of the mineral particle by crushing and grinding is essential to get a specific particle size for carbonation reaction. Iron ore (Fe_3O_4) and H_2O are separated from the mineral through magnetic separation step and heat treatment step respectively. Then this pre-treated mineral mix with water to form slurry. Carbonation reaction of this mineral slurry with compressed CO_2 is carried out in a carbonation reactor. After carbonation CO_2 and H_2O is recycled back to the reactor and carbonated product and by product is separated.

In recent years a lot of researches have been performed related to the different pathways of carbonation technology that can significantly reduce the emissions of CO_2 into the atmosphere from any carbon burning processes.

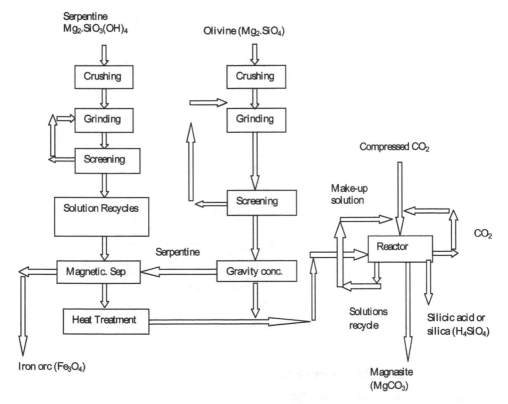

Fig. 12. Process flow diagram (redrawn) of carbonation process for magnesium silicate (O'Connor *et al.*, 2001)

Currently CO_2 capture processes are commercially used in the petroleum and petrochemical industries, but integration of CO_2 capture, transport and storage has not been demonstrated in a power plant yet, though several research and development programs have been demonstrated worldwide for power plants (IPCC, 2005). The status of different CCS technologies is given in Table 3.

CCS Component	CCS Technology	Research Phase	Demonstration phase	Economically feasible under specific conditions	Mature market
Capture	Post combustion			X	
	Pre combustion			X	
	Oxy-fuel combustion		X		
	Industrial separation (natural gas processing, ammonia production)				X
Transportation	Pipe line				X
	Shipping			X	
Geologic storage	Enhanced oil recovery (EOR)				X
	Gas or oil fields			X	
	Saline formations			X	
	Enhanced coal based methane recovery (ECBM)		X		
Ocean storage	Direct injection (dissolution type)	X			
	Direct injection (lake type)	X			
Mineral carbonation	Natural silicate minerals	X			
	Waste materials		X		
Industrial uses of CO_2					X

Table 3. Status of CO_2 capture and storage technology (IPCC, 2005)

It is clearly seen from Table 3 that mineral carbonation is still in research and development phase. Further study is essential. More in-depth review on mineral carbonation is presented in the following section.

4. Details of carbonation technologies

4.1 Routes or pathways of carbonation reaction

The carbonation reaction is carried out in two ways: direct routes and indirect routes. In direct routes carbonation takes place in a single step process, a gas-solid or a gas-liquid-solid process, and in indirect routes extraction of Ca/Mg occurs from the mineral and then carbonation takes place in an another step where silicates are directly carbonated in an aqueous medium at elevated temperature and CO_2 pressure (Huijgen, 2007). The detailed of pathways of carbonation reaction are described below.

4.1.1 Direct carbonation

4.1.1.1 Direct gas solid carbonation routes or dry carbonation

This process route with Ca/Mg-Silicates has some advantage for its direct reaction, simple process design, exothermic energy generated by the carbonation reaction that can be utilized in another process and no extra solvent is required to enhance the reaction (Lackner *et al.*, 1995; Huijgen, 2007). Figure 13 shows the direct gas solid carbonation process flowchart. Basic chemical reaction of this process is given below,

$$Ca/Mg\text{-}silicate\ (s) + CO_2\ (g) \rightarrow (Ca/Mg)\ CO_3\ (s) + SiO_2\ (s) + Heat$$

for instance in the case of olivine:

$$(Olivine)Mg_2SiO_4\ (s) + 2\ CO_2\ (g) \rightarrow 2\ MgCO_3\ (s) + SiO_2\ (s) + Heat$$

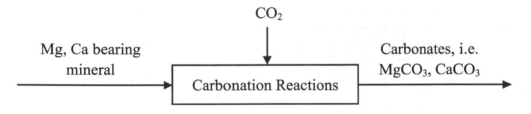

Fig. 13. Direct gas-solid carbonation process flowchart

Direct dry carbonation process occurs quite slowly at room temperatures and process can be accelerated by increasing the temperature to a certain degree (Lackner *et al.*, 1995). (Prigiobbe *et al.*, 2009) conducted carbonation tests of CO_2 with calcium rich fly-ash and found that the maximum conversions are occurred at 400°C temperatures or above 400°C. It was found from the test that the quickest carbonation ($t_{50\%}$ = 31 seconds) is occurred at 400°C with CO_2 occupying 50% of the flue gas and the maximum conversion (78.9%) occurred at 450°C with CO_2 occupying 10% of the flue gas. Though this process can result in high exothermic heat effect, but it is very energy-consuming due to the activation process of the mineral by heat treatment to enhance the carbonation rate significantly (Zevenhoven & Kohlmann, 2002).The reactant feed stocks are required to grind sufficiently to increase the surface area before carbonation reaction to ensure the effectiveness of the direct carbonation route.

4.1.1.2 Direct carbonation: Direct aqueous carbonation

According to the principle of carbonation process in natural weathering, the presence of water can enhance the carbonation reaction rate. Direct aqueous carbonation route has been developed based on the natural process of carbonation, as natural carbonation process is enhanced in presence of water (O'Connor *et al.*, 2000). Figure 14 shows the flow chart of direct aqueous carbonation process. The most attractive and promising route of carbonation is direct aqueous carbonation. A few solvents, such as sodium hydrogen carbonate ($NaHCO_3$), sodium chloride (NaCl), mg-acetate and water can be used as reaction medium to enhance the carbonation process. The purpose of this additive is to increase the dissolution and carbonation rate (Krevor, 2009; O'Connor *et al.*, 2005). This solvent can increases the carbonation process than direct dry carbonation process.

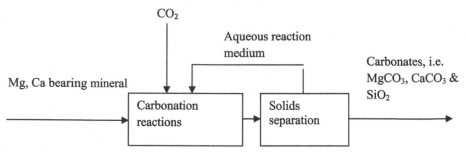

Fig. 14. Flow chart of direct aqueous carbonation process

Carbonation reaction can be increased by increasing the specific surface area, reducing (Ca, Mg)$^{2+}$-activity in solution and removing the SiO_2-layer. Addition of NaCl, $NaHCO_3$, and (Na/K)NO_3 can reduce the (Ca,Mg)$^{2+}$-activity in solution and increase the ionic strength. Addition of HCl, acetic acid, and citric acid can increase the reaction rate. In the case of serpentine, orthophosphoric acid, oxalic acid and EDTA can also increase the reaction rate (Park *et al.*, 2003).

4.1.2 Indirect carbonation: Cg/Mg-hydroxides carbonation or aqueous acid carbonation

The Cg/Mg -Hydroxides Carbonation or Indirect Carbonation or Aqueous Acid Carbonation essentially uses an aqueous solution to extract the magnesium and calcium content from the mineral prior to the actual carbonation process and the original mineral is split into smaller components, as a result carbonation take places at a much faster and favorable rate (Lackner *et al.*, 1995). Figure 15 shows the flowchart of aqueous acid carbonation process. Carbonation of Ca/Mg-Hydroxides or Indirect carbonation is considered as a feasible and faster process than carbonation of Ca/Mg-silicates, but high temperature and high CO_2 pressure are required for optimum conversion of mineral. It has been observed that 500°C and 340 bar CO_2 is required for $Mg(OH)_2$ carbonation for 100% conversion within two hour (Lackner *et al.*, 1997). The chemical reaction of the carbonation of $Mg(OH)_2$ is shown below,

$$Mg(OH)_2 \;(s) \rightarrow MgO \;(s) + H_2O \;(g) \;(dehydroxylation)$$

$$MgO \;(s) + CO_2 \;(g) \rightarrow MgCO_3 \;(s) \;(carbonation)+ Heat$$

The activation energy of the combined carbonation and dehydroxylation reaction are 304 kJ/mol and the optimum temperature for the reaction is 375 °C which is thermodynamically favourable (Butt *et al.*, 1996; Huijgen, 2007). Though, Ca/Mg-hydroxides are not available in nature, but Ca/Mg-hydroxides can be found by the conversion of Ca/Mg-Silicates using extraction agent (Lackner, 1995; Huijgen, 2007).

Depending on the mineral and characteristics of the overall process, various acids, bases or water such as hydrochloric acid, caustic soda, sulphuric acid or steam can be added at various stages to extract the Ca/Mg components from the magnesium/calcium bearing mineral. Due to the losses of the extraction agent throughout the process hydrochloric acid is perhaps a preferred choice due to its relatively low cost. Recovery of the extraction agent at the end of the extraction process is essential for the overall feasibility of aqueous acid carbonation process (Lackner *et al.*, 1995). Literature indicates that several extraction agents can be used to extract Ca/ Mg component from mineral such as, HCl, molten salt (MgCl₂.3.5H₂O) and acetic acid.

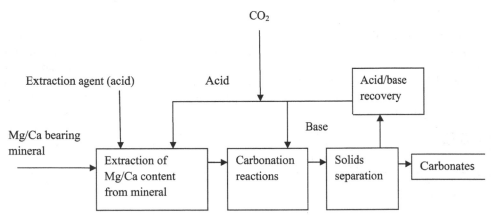

Fig. 15. Flowchart of aqueous acid carbonation process

4.2 Materials for carbonation

Mineral CO₂ sequestration requires large amounts of Ca/Mg minerals for large-scale sequestration of CO₂ from flue gases. For example 1 kg of CO₂ may require 2 kg (or more) of serpentine for the carbonation reaction, which may cause a significant environmental impact due to the disposal problem (Sipila *et al.*, 2007). About 10,000 tons of CO₂/day is produced from a single 500 MW power station. To sequester this large amount of CO₂ through the carbonation process, approximately 23,000 tons/day of magnesium silicate ore is required (Lackner *et al.*, 1995). Different types of materials can be used for the carbonation process such as mineral rocks and industrial residues. Calcium and magnesium oxides and hydroxides are the ideal materials for carbonation, but they are not available in nature. Alkaline mineral materials are available and abundant in natural silicate rocks around the world that contain high amounts of magnesium, calcium and also low amounts of iron, sodium and potassium.

Table 4 shows the composition of some selected rocks and pure minerals and their potential CO₂ sequestration capacity. Industrial residues such as slag from steel production, fly ash from coal combustion, de-inking ash from recycling of paper and municipal residue can also be used for mineral carbonation to minimise CO₂ emissions (IPCC, 2005; Johnson, 2000).

These industrial residues are a valuable source of Ca/Mg, therefore industrial residues can be applied to carbonation processes to minimise CO₂ emissions. This has a great

Rock mineral	MgO[wt%]	CaO[wt%]	RCO$_2$[kg/kg]^a
Serpentinite	~40	~0	~2.3
Serpentinite, Mg$_3$Si$_2$O$_5$(OH)$_4$	48.6		1.9
Dunite	49.5	0.3	1.8
Olivine, Mg$_2$SiO$_4$	57.3		1.6
Wollastonite	0.8	43.7	2.9
Wollastonite, CaSiO$_3$		48.3	2.6
Talc	34.7	0.0	2.6
Talc Mg$_3$Si$_4$O$_{10}$(OH)$_2$	31.9		2.9
Basalt	6.2	9.4	7.1

a, RCO$_2$ = mass ratio of rock to CO$_2$ required for CO$_2$ sequestration

Table 4. Composition of some selected rocks and pure minerals and their potential CO$_2$ sequestration capacity (Huijgen, 2007)

environmental benefit, as these residues are disposed of easily and cost effectively by reusing them in carbonation technology. Table 5 shows different studies on carbonation of industrial residues, described below. Industrial residues have some potential benefit compared to mineral materials as they are readily available and cheap, and also utilising these residues in carbonation processes has a positive impact on the environment. Besides this, due to chemical instability, industrial residues are highly reactive in nature.

Residue	Details of carbonation process	T [°C]	P$_{CO2}$ [bar]	d [mm]
Blast furnace slag		25	3	NA^a
Cement-immobilised slag	Supercritical CO$_2$	50	250	
Coal fly ash	Aqueous route ^{b}	185	115	NA
Coal fly ash	20% moisture	25	2.8	<0.25
De-inking ash		25	3	NA
FBC coal ash	Aqueous route ^{b}	155	75	NA
FGD coal ash	Aqueous route ^{b}	185	115	NA
MSWI ash		25	3	NA
MSWI bottom ash	Water content: moisture.	20	60(l)	<10
		40	150(sc)	
		50	250(sc)	
OPC cement		25	3	NA
Portland cement pastes	W/c: 0.6	59	97	NA
Pulverised fuel ash	Water:solid 0.1 – 0.2	25	3	NA
Spent oil shale	20% moisture	25	2.8	<0.25
Stainless steel slag		25	3	NA
Waste Dravo-Lime	Aqueous route ^{b}	185	115	NA

^{a} Not available. ^{b} Additives used: 0.5M Na2CO3 / 0.5M NaHCO3 / 1.0M NaCl.

Table 5. Different studies on carbonation of residues with process condition (Huijgen, 2007)

4.3 Pre-treatment before carbonation

Mineral carbonation requires some pre-processing of minerals such as grinding, heat treatment and magnetite separation, and also the CO_2 pressure should be similar to the pressure of the pipeline if CO_2 is piped to the disposal site (O'Connor et al., 2001). Before carbonation, CO_2 is preheated typically between 100°C and 150°C, especially for aqueous carbonation processes where the carbonation occurs between 300°C to 500°C (Butt et al., 1996). Mineral treatment with steam or acid or a combination of these can be applied to increase the specific surface area of a mineral to improve the carbonation rate (O'Connor et al., 2001). Carbonation has been successfully performed after such activation, heat treatment, grinding and magnetite separation.

Mineral and carbonation reaction	T_{deh} k	T_{max} K	ΔH kJ/mole	ΔQ kJ/mole
Calcium Oxide CaO + $CO_2 \to CaCO_3$	-	1161	-167	87
Magnesium Oxide MgO + $CO_2 \to MgCO_3$	-	680	-115	34
Calcium Hydroxide $Ca(OH)_2 + CO_2 \to CaCO_3 + H_2O$	791	1161	-68	114
Magnesium Hydroxide $Mg(OH)_2 + CO_2 \to MgCO_3 + H_2O$	538	680	-37	46
Wollastonite $CaSiO_3 + CO_2 \to CaCO_3 + SiO_2$	-	554	-87	37
Clinoenstatite (Pyroxene) $MgSiO_3 + CO_2 \to MgCO_3 + SiO_2$	-	474	-81	23
Forsterite (Olivine) $1/2Mg_2SiO_4 + CO_2 \to MgCO_3 + 1/2SiO_2$	-	515	-88	24
Diopside (Pyroxene) $1/2CaMg(SiO_3)_2 + CO_2 \to 1/2CaCO_3 + 1/2MgCO_3 + SiO_2$	-	437	-71	19
Grossular (Garnet) $1/3Ca_3Al_2Si_3O_{12} + CO_2 \to CaCO_3 + 1/3Al_2O_3 + SiO_2$	-	465	-65	28
Anorthite (Feldspar) $CaAl_2Si_2O_8 + CO_2 \to CaCO_3 + Al_2O_3 + 2SiO_2$	-	438	-81	39
Anorthite Glass $CaAl_2Si_2O_8 + CO_2 \to CaCO_3 + Al_2O_3 + 2SiO_2$	-	691	-148	121
Pyrope (Garnet) $1/3Mg_3Al_2Si_3O_{12} + CO_2 \to MgCO_3 + 1/3 Al_2O_3 + SiO_2$	-	533	-92	40
Talc $1/3Mg_3Si_4O_{10}(OH)_2 + CO_2 \to MgCO_3 + 4/3SiO_2 + 1/3H_2O$	712	474	-44	64
Tremolite (Amphibole) $1/7Ca_2Mg_5Si_8O_{22}(OH)_2 + CO_2 \to 2/7 CaCO_3 + 5/7MgCO_3 + 8/7 SiO_2 + 1/7H_2O$	839	437	-37	72
Chrysotile (Serpentine) $1/3Mg_3Si_2O_5(OH)_4 + CO_2 \to MgCO_3 + 2/3SiO_2 + 3H_2O$	808	680	-35	78

[NOTES:
T_{max}, is the maximum carbonation temperature for $PCO_2 = 1$ bar.
T_{deh} refers to the dehydroxylation temperature,
The enthalpy of reaction, $\Delta H(T_{max})$, is normalised to one mole of CO_2
The heat ΔQ is the energy required to heat the original mineral and CO_2 to the higher of T_{max} and T_{deh}, normalised to one mole of CO_2.
The initial temperature is assumed to be 298K.] (Lackner et al., 1995)

Table 6. Thermodynamic properties of carbonation reactions (Lackner et al, 1995)

Particle size is a vital factor for determining carbonation reaction rates, as carbonation reactions are mainly surface controlled. Particles with coarser sizes can be used, but smaller particle sizes are preferred for carbonation reaction. Particle sizes < 37 microns are preferred for optimum reaction. Particle size can be reduced through grinding and crushing. An increase from 10 to 90% carbonation conversion can be achieved by reducing the particle size from 106-150 μm to <37 μm (O'Connor et al., 2000).

Before carbonation heat treatment is required to remove chemically bound water and activates the mineral for carbonation process. It was roughly estimated that the optimum energy requirements for heat treatment is 200 kWh/ton for serpentine at 600- 650ºC it was found that the specific surface area of antigorite is increased from 8.5 to 18.7 m^2/g by heat treatment to accelerated the reaction(O'Connor et al., 2001). Heat treat process generates steam which is also a potential recoverable energy. Magnetic separation step is mainly used for serpentine to remove the magnetite (Fe_3O_4) that is remaining naturally with the serpentine. Magnetic is a useful by-product.

The natural process of mineral carbonation is very slow at ambient temperatures. It has been evaluated that carbonation reaction rates can be increased by increasing the temperature and pressure (Lackner et al., 1995; Zevenhoven & Kohlmann, 2002). The maximum allowable carbonation temperature should be maintained during carbonation reaction according to the nature of the mineral and the CO_2 pressure. Above this temperature, carbonation reactions are not thermodynamically feasible. Table 6 shows the thermodynamic properties of various carbonation reactions.

As carbonation reaction is exothermic, carbonated product can be obtained at low temperature. Above 900°c for calcium carbonate and above 300°c for magnesium carbonate with CO_2 partial pressure of one bar the carbonation reaction can be reversed that is called calcinations (IPCC 2005; Nikulshina et al., 2007).

4.4 Energy consumption and exothermic nature of carbonation processes

Carbonation processes require energy intensive preparation of the solid reactants including pre-treatment, mining, transportation, grinding and activation to achieve the required conversion, and power plant efficiency can be reduced by this. Power plants' efficiency can be reduced by up to 20 percent for CO_2 capturing processes (Plasynski and Chen, 2000), though chemical reaction of the mineral carbonation process is exothermic. If suitable energy recovery options are implemented, then operating efficiency of those power plants will be raised using the exothermic energy and energy recovered from the product. The wet carbonation of the natural silicate olivine causes a 30-50% energy penalty on the power plant, and to implement a CCS system in such power plants, 60-180% more energy is required to reduce CO_2 emission while maintaining equivalent power output (IPCC, 2005). It is estimated that efficiency is reduced 27% by implementing mineral CO_2 sequestration technology in power plants and 75% of the total efficiency reduction is caused by grinding. Energy cost can be reduced by using the exothermic nature of the reaction and also utilising the by-products of the carbonation reaction. Furthermore, reuse of the resulting products could enhance the economic return (Huijgen, 2007).

It was estimated that total costs of CO_2 capture by mineral carbonation are about 90-120 €/tonCO_2 which is very high for large scale CO_2 capture by mineral carbonation technology, though it was observed that costs can be reduced using the exothermic nature of the carbonation reaction and also utilising the valuable by-product of mineral mining such as magnesium, silicon, chromium, nickel and manganese and the reusing of carbonated product (Huijgen & Comans, 2003). Some possibilities for energy integration by exothermic reaction of carbonation technology have been identified using Aspen Plus modelling software. These are listed in Table 7 below (Brent and Petrie, 2008)

Energy Sinks	Energy Sources
Mineral comminution energy	Compressed CO_2 product stream
CO_2 compression duty	Dehydroxylation reaction products
Serpentine activation	Carbonation reaction products
Carbonation reagent slurry preheating	

Table 7. Possible Energy sinks and energy sources (Brent and Petrie, 2008)

5. CCS projects in Australia

Currently, several carbon capture and storage (CCS) projects are running around across the world including Australia. Some CCS projects in Australia are detailed below,

5.1 Zero Gen Project

Stanwell Corporation (owned by the Queensland Government) proposes to build a 100 MW Integrated Gasification Combined Cycle (IGCC) plant with capture technology adjacent to the existing Stanwell Power Station, 29 kilometres west of Rockhampton. The ZeroGen project will integrate IGCC technologies with Carbon Capture and Storage (CCS) to produce low-emission base load electricity. The project will convert pulverised coal into a synthesis gas (consisting mainly of hydrogen and carbon monoxide) through the IGCC process with a mixture of oxygen enriched air and steam under pressure. The "syngas" then undergoes a shift conversion in a gasifier where the carbon monoxide is converted to hydrogen and carbon dioxide. The CO_2 is then separated from the shifted "syngas" and produces a clean, low-carbon, high-hydrogen fuel to produce electricity. The estimated project cost is A\$4.3 billion. It is expected that this will have the potential to capture up to 90% of CO_2 emissions for full sequestration. Japan-based Mitsubishi Corporation (MC) and Mitsubishi Heavy Industries (MHI) will provide IGCC and carbon capture technology for the power plant. The proposed plant will be operational in 2015 (Zero Gen Project, 2008).

5.2 Callide Oxyfuel Project

The Callide Oxyfuel Project is located at Biloela in Queensland and aims to capture carbon using oxyfuel combustion, combined with carbon storage. In this technology, coal is combusted in a boiler with oxygen and then exhaust gases have been recycled instead of regular air. The project is lead by CS Energy Ltd with international partners IHI Corporation (Japan), J-Power (Japan), Mitsui & Company (Japan), Schlumberger Oilfields Australia and Xstrata Coal. Financial support for this project is provided by the Australian Coal Association and the Commonwealth, Queensland and Japanese governments. Stage one of the project involves the conversion of a generator to apply oxyfuel combustion and the capture of CO_2. Stage two of the project will see the 'transport, injection and storage of liquefied CO_2 in deep geological formations'. The Callide Oxyfuel project will retrofit the 30

MW generators at the Callide "A" pulverised coal power station near Biloela in Queensland to allow oxyfuel combustion (Callide Oxyfuel Project, 2008).

5.3 Delta-Munmorah Post Combustion Capture (PCC) Project
Delta-Munmorah PCC Project, run by the CSIRO with Delta Electricity, is based on post-combustion (where CO_2 is captured after combustion) CO_2 capture at an ammonia based pilot plant built at the Munmorah Power Station on the New South Wales Central Coast. This pilot plant research project began in February 2009 and will continue until 2013. This project is also supported by the Australian Government (National Research Flagships, CSIRO, 2009).

5.4 Wandoan Power Project
The Wandoan Power Project will integrate Gasification Combined Cycle (IGCC) pre-combustion carbon capture and storage technologies. In IGCC process fossil fuel is first converted into CO_2 and Hydrogen gas (H_2). Then CO_2 gas is separated and electricity is produced by the combustion of Hydrogen-rich gas. This project is led by GE Energy, with partners Stanwell Corporation and Xstrata Coal. This project is located near Wandoan in Queensland's Surat Basin and will build a 400MW IGCC power plant with carbon capture and storage capacity of 90% of CO_2 emissions (Flagship Project: Wandoan Project).

5.5 Project on mineral carbonation at NSW
Recently a joint venture project between Green Mag Group and the University of Newcastle, Australia has been established with an aim to set a carbonation plant at Newcastle, Australia using serpentine that is abundant in Australia. In this study, the efficacy and prospect of carbonation technology (Mineral carbonation project for NSW) will be investigated and evaluated.

6. Comparison of CCS technologies

As mentioned earlier, there are many CO_2 capture technologies available to mitigate CO_2 emissions such as chemical absorption, pressure swing adsorption (PSA), gas separation membranes and cryogenic separation. It can be noted that implementation of any CCS technology will introduce extra cost and energy penalty. So cost and energy penalty should be considered before implementation of CCS technology in power plant. Different technologies are compared with their advantages and disadvantages in Table 8.

Technology	Advantage	Disadvantage
Chemical Absorption Technology	• Technically mature • Suitable for low concentration of gas stream • Solvent can be regenerated • Purity of the CO_2 stream up to 95%	• High energy is required for solvent regeneration and also high capital cost associated with absorber and stripper section • Large amount of solvent is required to capture CO_2 • Up to 20% to 30% energy is required for CO_2 capture process through chemical

Technology	Advantage	Disadvantage
		absorption out of the total output energy of a 500 MW power plant • Absorption capacity of solvent is limited
Physical Absorption Technology	• Technically mature and common processing equipment is used • Less energy is required to regenerate the solvent compared to chemical absorption • Flue gas containing NOx, O$_2$ and CO can pass through the physical absorption process as they do not degrade the solvent due to their low solubility in the physical solvent	• Physical solvent is costly • Suitable only for high pressure gas streams • High capital cost due to absorber, stripper and solvent
Membrane Technology	• Suited to high pressure gas streams with low concentrations	• Technically immature • Energy intensive • High equipment cost due to compressors, membrane housing, membrane filters • High pressure differential is required between two phase
Adsorption Process	• Suitable for low concentration streams	• High energy consumption and high capital cost for adsorbed bed and sorbents • Low efficient
Low Temperature Systems	• Most suitable for binary gas streams • CO$_2$ product is ready for transport as CO$_2$ is separated in liquid form so there is no requirement for compression before transport • Separated CO$_2$ stream purity relatively high	• Energy and cost intensive due to refrigeration and distillation units • Efficiency affected by other gas components in flue stream • Traces elements of flue gases (such as methane or water vapour) should be required to be removed before cryogenic separation (Ho, 2007)
Mineral Carbonation Technology	• Availability of feedstock • Exothermic nature • Valuable by-product	• Energy consuming process, especially for pre-treatment

(Ho, 2007).

Table 8. Comparison of different CO$_2$ capture technologies

After a careful consideration of the advantages and disadvantages of existing CO_2 removal technologies, it can be said that mineral carbonation technology is a potentially viable CO_2 reduction technology for a power plant, mainly due to its exothermic nature which can be integrated with other energy consuming requirements of the plant provided a suitable energy recovery process is applied in the power plant. This technology is relatively new and promising, and still in the research phase. So further research on carbonation technology is required because of the following advantages compared to other technologies:

- Carbonated products are environmentally safe and stable over geological time frames.
- Raw materials for mineral carbonation exist in vast quantities across the world.
- Carbonation processes are exothermic.
- Carbonation reactions produce value-added carbonated products which have some valuable uses such as mine reclamation, construction or other prospective applications.
- Carbonation reaction and mining of mineral produce some valuable by products such as chromium, nickel, magnesium, silicon, and manganese which have valuable market use.
- This mitigation option is feasible where underground reservoirs and ocean storage of CO_2 is not possible.

7. Conclusions

It can be concluded that the improvement of cost and energy penalty associated by carbonation technology can be achieved through integration of exothermic energy produced from the carbonation reaction, energy recovered from product of carbonation reaction for self sustaining system and utilizing the carbonated product and by-product. Another prospect of this technology is utilizing the industrial waste as a feed stock to make this process less expensive. Any other CCS technologies have not these promising prospects. There are studies available on acceleration of the carbonation processes related to the reaction kinetics, but there is very limited information related to the integration of the carbonation processes into power plants, and the performance of the power plants associated with this process. To integrate carbonation processes into coal fired power plants, research should be done to assess the performance of these processes to introduce this technology at an affordable rate with reduced energy penalties to power plants. The outcome of the study may contribute to the savings of process energy, hence decreasing processing costs and increasing capital gain.

8. References

ABS, (2001). Energy and greenhouse gas emissions accounts, Australia, Australian Bureau of Statistics, Canberra, Report No. 4604.

AGO, (2006). National Greenhouse Gas Inventory, Australian Greenhouse Office, Canberra. (Available at www.greenhouseoffice.gov.au)

Bhide, B. D., Voskericyan, A. and Stern, S. A. (1998). Hybrid processes for the removal of acid gases from natural gases, *Journal of Membrane Science*, (140), 27-49.

Brent, G. F. and Petrie, J.G. (2008). CO2 sequestration by mineral carbonation in the Australian context, Chemeca 2008: Towards a Sustainable Australasia.

Butt, D.P., Lackner, K.S., Wendt, C.H., Conzone, S.D., Kung, H., Lu, Y.-C. & Bremser, J.K. (1996). Kinetics of thermal dehydroxylation and carbonation of magnesium hydroxide, *Journal of American Ceramic Society* 79(7), 1892-1898.

Callide oxy fuel project, viewed at 18 July, 2010.
http://www.callideoxyfuel.com/Why/OxyfuelTechnology/tabid/72/Default.aspx

Coal Industry Advisory Board. (1994). Industry attitudes to combined cycle clean coal technologies survey of current status, Organization for Economic Co-operation and Development / International Energy Agency, France.

Chue, K. T., Kim, J. N., Yoo, Y. J., Cho, S. H. and Yang, R. T. (1995). Comparison of activated carbon and zeolite 13X for CO$_2$ recovery from flue gas by pressure swing adsorption, *Industrial & Engineering Chemistry Research,* 34 (2), 591-598.

Cullinane, J.T., Oyenekan, B.A., Lu, J, Rochelle, G. T. (2002). Carbon Dioxide Capture by Absorption with Potassium Carbonate, NETL funded project, University of Texas at Austin.

Demirbas, A. (2005). Potential applications of renewable energy sources, biomass combustion problems in boiler power systems and combustion related environmental issues. Progress in Energy and Combustion Science, Volume 31, Issue 2, 2005, Pages 171-192.

Demirbas, A. (2008). Carbon Dioxide Emissions and Carbonation Sensors Turkey, *Energy Sources,* Part A, 30:70–78.

Dindore, V. Y., Brilman, D.W.F., Feron, G.F.H. and Versteeg, G.F. (2004). CO$_2$ absorption at elevated pressures using a hollow fiber membrane contactor, Sep. Purif. Technol. 40, 133–145

DOW, (2005) Speciality Alkanolamines: Product information, The Dow Chemical Company (Herzog & Golomb, 2004).

Department of Energy and Environmental Protection Agency, Washington DC, 2000).

ESAA, (2003). *Electricity Australia 2003,* Electricity Supply Association of Australia, Sydney.

Flagship Project: Wandoan Project, viewed at 18 July, 2010,
http://www.newgencoal.com.au/active-projects_flagship-projects.aspx?view=13

Fei, W. (2004). Plum Flower Mini Ring – A highly efficient packing for CO$_2$ capture and separation, In Proceedings of the Separations Technology VI: New Perspectives on Very Large- Scale Operations, Fraser Island.

Fei, W. Y. and Song, X. Y. (2005) Comparison of packings for absorption with high liquid loading, AIChE Spring National Meeting, Conference Proceedings; AIChE Spring National Meeting, Conference Proceedings, 1817.

Feron, P. H. M. (1992). CO$_2$ capture: The characterisation of gas separation/removal membrane systems applied to the treatment of flue gases arising from power generation using fossil fuel, Cheltenham, IEA/92/08, IEA Greenhouse Gas R & D Programme.

Goldthorpe S. H., Cross, P. J. I., and Davison J. E. (1992). Systern Stidies on CO$_2$ abatement from Power Plants; *Energy Conversion Management.* Volume 33, Issues 5-8, pp 459-466.

GPSA, (2004). Engineering Data Handbook, Gas Processors Suppliers Association, Tulsa.
http://www.geos.ed.ac.uk/sccs/capture/postcombustion.html

Ho, M.T. (2007). Techno-economic modeling of CO$_2$ captures systems for Australian industrial sources, PhD Thesis.

Herzog, H., Golomb, D. and Zemba, S. (1991). Feasibility, modeling and economics of sequestering power plant CO_2 emissions in the deep ocean, *Environmental Progress*, 10(1), 64-74.

Huijgen, W.J.J & Comans, R.N.J. (2003), Carbon dioxide sequestration by mineral carbonation, Literature Review, ECN-C--03-016.

Huijgen, W. J. J. (2007). Carbon Dioxide Sequestration by Mineral Carbonation, PhD Thesis.

IPCC, (2005). Carbon Dioxide Capture and Storage, Cambridge University Press, UK. pp 431.

International energy agency, (2004). Prospects for CO_2 capture and storage, OECD/IEA.

International Energy Agency - Greenhouse Gas R&D Programme (IEA-GHG), (2000). CO_2 storage as carbonate minerals, prepared by CSMA Consultants Ltd, PH3/17, Cheltenham, United Kingdom

Johnson, D.C. (2000). A solution for carbon dioxide overload, *SCI Lect. Pap. Ser.* 108 (2000), pp. 1–10.).

Kojima, T., Nagamine, A., Ueno, N. & Uemiya, S. (1997). Absorption and fixation of carbon dioxide by rock weathering, *Energy Conversion and Management*, 38, S461-466.

Krevor, SC & Lackner, K.S. (2009). Enhancing process kinetics for mineral carbon sequestration, *Energy procedia*, vol.1, pp. 4867-4871.

Kreutz, T.G., Williams, R.H., Socolow, R.H., Chiesa, P., and Lozza, G. (2002). Production of Hydrogen and Electricity from Coal with CO_2 Capture, Proceedings of the 6th International Conference on Greenhouse Gas Control Technologies (GHGT6), J. Gale and Y. Kaya (editors), Kyoto, Japan.

Lackner, K.S., Wendt, C.H., Butt, D.P. Joyce, E.L. & Sharp, D.H. (1995). Carbon dioxide disposal in carbonate minerals, *Energy* 20(*11*), 1153-1170.

Lackner, K.S., Butt, D.P. & Wendt, C.H.. (1997). Progress on binding CO2 in mineral substrates, *Energy Conversion and Managemen,t* 38, S259-264.

Lu, J. G., Zheng, Y.F. and Cheng, M. D.(2009). Membrane contactor for CO_2 absorption applying amino-acid salt solutions, Desalination 249, 498–502.

Lu, J. G., Wang, L. J., Sun, X..Y., Li, J. S., and. Liu, X. D. (2005). Absorption of CO_2 into aqueous solutions of methyldiethanolamine and activated methyldiethanolamine from a gas mixture in a hollow fiber contactor, *Ind. Eng. Chem. Res.* 44, 9230–9238.

Leci, C. L. (1996). Financial implications on power generation costs resulting from the parasitic effect of CO2 capture using liquid scrubbing technology from power station flue gases, *Energy Conversion and Management*, 37 (6-8), 915-921.

Mimura, T., Simayoshi, H., Suda, T., Iijima, M. and Mituoka, S. (1997) Development of energy saving technology for flue gas carbon dioxide recovery in power plant by chemical absorption method and steam system, *Energy Conversion and Management*, 38 (Supplement 1), S57-S62.

MIT, (2007). The Future of Coal, p. 35.

Marchetti, C. (1977). On geo engineering and the CO_2 problem, Climatic Change, 1, 59-68.

Mineral carbonation project for NSW.*http://www.sustainabilitymatters.net.au/articles/41409-Mineral-carbonation-project-for-NSW*

National research flagships, CSIRO, 2009, Post-combustion capture at Munmorah http://www.de.com.au/Sustainability/Greenhouse/Carbon-Capture-Research-Project/Carbon-Capture-Research-Project/default.aspx, Viewed at 15July, 2010.

Nikulshina, V., Galvez, E. and Steinfeld. (2007) A. kinetic analysis of the carbonation reactions for the capture of co₂ from air via the ca(oh)₂–caco₃–cao solar thermo chemical cycle, *chem. eng. j.* 129, pp. 75–83.

Narula, R.G., Wen, H. and Himes, K. (2002). Incremental cost of CO₂reduction in power plants, Proceedings of IGTI ASME TURBO EXPO.

O'Connor, W.K., Dahlin, D.C., Nilsen, D.N., Rush, G.E., Walters, P R. and Turner, P.C. (2001). Carbon Dioxide Sequestration by Direct Mineral Carbonation: Results from Recent Studies and Current Status, presented at the First National Conference on Carbon Sequestration, Washington, DC.

O'Connor, W.K., Dahlin, D.C., Nilsen, D.N., Walters, R.P. & Turner, P.C. (2000) Carbon dioxide sequestration by direct mineral carbonation with carbonic acid, 25th International Technical Conference on Coal Utilization and Fuel Systems, Clearwater, FL, USA.

O'Connor, W.K., Dahlin, D.C., Rush, G.E., Gerdemann, S.J., Penner, L.R. & Nilsen, D.N. (2005) Aqueous mineral carbonation: mineral availability, pretreatment, reaction parametrics and process studies, DOE/ARC-TR-04-002, Albany Research Center, Albany, OR, USA.

Plasynski, S.I. and Chen, .Z.Y. (2000). Review of carbon capture technology and some improvement opportunities, American Chemical Society, 220 (2000) U391-U391.

Park, A.-H.A., Jadhav, R. & Fan, L.-S. (2003) CO₂ mineral sequestration: chemically enhanced aqueous carbonation of serpentine, *Canadian Journal of Chemical Engineering* 81(3), 885-890.

Parker, L., Folger, P. and Stine, D.D. (2008) Capturing CO₂ from Coal-Fired Power Plants: Challenges for a Comprehensive Strategy, CRS report for congress.

Prigiobbe, V., Polettini, A. and Baciocchi, R.. (2009), Gas–solid carbonation kinetics of Air Pollution Control residues for CO₂ storage, Chemical Engineering Journal, Volume 148, Issues 2-3, Pages 270-278.

Powerspan Corp. (2008). Carbon Capture Technology for Existing and New Coal-Fired Power Plants.

Ryan M. D. and Donald S. S. (2008), An Introduction to CO₂ Capture and Sequestration Technology, *Utility Engineering*, p. 7.

Sipila, J, Teir, S and Zevenhoven, R..(2007). Carbon dioxide sequestration by mineral carbonation Literature review update, 2005–2007, report VT 2008-1.

Seifritz, W. (1990), CO₂ disposal by means of silicate, *Nature*, vol. 345, pp. 48.

Skarstrom, C. W. (1960). Method and apparatus for fractionating gaseous mixtures by adsorption, Patent. No. 2944627, US.

Scottish carbon capture and storage, viewed at 18 July, 2010
http://www.geos.ed.ac.uk/sccs/capture/precombustion.html

Scottish carbon capture and storage, viewed at 18 July, 2010
http://www.geos.ed.ac.uk/ sccs/capture/oxyfuel.html

Scottish carbon capture and storage, viewed at 18 July, 2010
http://www.geos.ed.ac.uk/sccs/capture/postcombustion.html

World coal institute, http://www.worldcoal.org/coal/uses-of-coal/coal-electricity , viewed at 20 July, 2010.

Williams, M., (2002). Climate change: information kit, Geneva: the United Nations Environment Programme (UNEP) and the United Nations Framework Convention on Climate Change (UNFCCC).

Williams, R.H. (2003). Decarburization of Fossil Fuels for the Production of Fuels and Electricity, presented at the Canadian Clean Coal Technology Roadmap Workshop, Calgary, Alberta.

Zevenhoven, R. & Kohlmann, J. (2002). CO_2 sequestration by magnesium silicate mineral carbonation in Finland, Recovery, *Recycling & Re-integration*, Geneva, Switzerland.

Zero Gen Project, http://www.zerogen.com.au/factsheets.aspx, viewed at 18 July, 2010.

11

Heat-Resistant Steels, Microstructure Evolution and Life Assessment in Power Plants

Zheng-Fei Hu
School of Materials Science and Engineering,
Tongji University, Shanghai
China

1. Introduction

Most parts of the electricity generating equipments in power plants work at elevated temperature and high steam pressure, including boiler, turbine and connected system of tubing and piping. The generating equipments operate with steam pressures in the range of 20MPa or even more and the steam temperature is also high in the range of 600 °C. This will be the condition of water and steam inside the circulates through generating system. On the furnace side in boiler, the gas temperatures outside of the tubes can be as high as 1400 °C. The water or steam carrying tubes, the drum and other connected parts have to be strong enough to withstand these temperatures and pressures. In power plant, great efficiency means a saving in fuel and less emission of carbon dioxide in a given electricity output, which consequentially reduces the rate at which damage is done to the globule environment. The efficiency of steam turbines can be obviously improved by increasing the maximum operating pressure and temperature. That is why there is a tendency that operating parameters become much higher to promote efficiency.

We can imagine what will happen if the tubes or pipes burst open at such high pressure and temperature, which will be a catastrophic accident and lead to fatalities. To prevent such failures every detail should be subject to strict adherence to code compliances, from the metallurgy of the steel used, strength calculations for each part to manufacturing and welding procedures. Every part is tested and results documented to ensure compatibility with the codes. At the end of the construction phase, Hydro tests at pressure of 1.5 to 2 times the operating pressure were carried to ensure that all parts can withstand the high pressure conditions. Under normal operating conditions, the pressure parts can withstand these high temperatures and pressures for years. If there is an abnormal increase in pressure, the safety valves will automatically open and let out the steam to the atmosphere.

Although safe design and careful condition monitoring have always been of great concern for the power industry, high temperature components loaded with steam pressure in power plants have a high damage potential during long-term service. Today this issue becomes more and more important with the higher and higher steam parameters. So a great number of standards, rules and guidelines exist worldwide to avoid catastrophic failures and to deliver the right basis for condition based inspection. Furthermore, technical advances in design, instrumentation, condition monitoring and materials are driving forces for the enhancement of life prediction and assessment procedures. However, well understanding of material degradation in service is the basis for improved inspection and scheduling maintenance to keep the plants in reliability, availability and profitability.

With the development of power technology and heat resistant steels, the later the power plant was constructed, the higher the operating parameters (temperature and steam pressure and gross generation). For the reason of developing history, today many different level power plants operate at different parameters around the world, and the generating components were manufactured from various alloys using a wide range of fabrication techniques. Because the design conditions of boilers vary within a broad power generating system, there are many different alloys being used in various product forms. New alloys with complex metallurgy have being introduced for decades, and the information base for well established alloys continues to grow.

It is believed that creep and material structure evolution are tightly related. Components subjected to creep stress have a limited lifetime. In fact, a large number of studies have been performed in order to relate microstructural investigation and service exposure or residual life. Many reports have involved this issue [1,2,3].

This chapter is intended to give an overview of the heat resistant steels used in power plant, including development of these materials, their application, techniques applied for microstructural investigation in order to assess residual life.

2. Heat resistant steels and application

Here it is presented a general idea of heat-resistant steels and a brief review of its current status for power generation applications. And some information on microstructural characteristics, appropriate for the application, data of critical properties are given for selected materials. We attempt to give an overall impression of the advantages and limitations of the various classes of steels with the emphasis on applications.

2.1 Profile of heat resistant steels

Generally, there are two fundamental classes of heat resistant steels used in power plants, the "ferritic/martensitic" and the "austenitic". The ferritic/martensitic steels have the same body-centered cubic crystal structure as iron. They are simply iron containing with relatively small addition alloy elements, such as the main element chromium added from 2% to about 13%. These ferritic/martensitic grades also have a little manganese, molybdenum, silicon, carbon and nitrogen, mostly included for their benefits in precipitation strengthening and prompting high temperature behavior. Ferritic grades are used broadly because they are much economic for their low content of alloy additions. They also have some resistance to oxidization at red heat, and which is in direct proportion to chromium content.

Ferritic/martensitic steels used in high temperature can be divided in two classes by microstructrue and content of additions. One class are usually named low alloy steels, which contain 1-3%Cr and total alloy elements less than 5%; another are called 9-12Cr martensitic steels and commonly contain alloy elements in the range 10-20%. With the operating parameters increasing in power plants, more and more components fabricated with martensitic steels, and pearlitic steels have a tendency to be replaced partly. In recent years, worldwide efforts to increase efficiency in power plants have created a demand for steels that can withstand higher pressure and higher service temperatures. The representative developments are grade P/T91 – X10CrMoVNb9-1, T/P92, and so on, which are modification of the existing 9-12Cr% grade with additions of vanadium, niobium, nitrogen et al. Till today, the most advanced martensitic steels exposes at the steam temperature less than 650 °C.

When enough nickel is added to the iron−chromium mix, the alloy becomes austenitic which has a face-centered cubic crystal structure. Austenitic steels have much highly strength and ductility and also have much greater creep-rupture strength than the ferritic/martensitic steels. At room temperature the austenitics are more ductile and generally easier to fabricate. Austenitic steels are much expensive for their high contents of alloy element additions, which are usually used at the steam temperature over 650°C.

2.2 Effects of alloy elements

Heat Resistant steels are those solid solution strengthened alloy steels for use at temperatures over 500°C and limited in the extreme to750°C. As they are used over a certain broad temperature ranges, these steels usually were strengthened by hard mechanism of heat treatment, solid solution and precipitation. In order to achieve the desired properties, all those heat-resistant steels are composed with several alloy elements except one or two basis elements. Their complicated compositions achieve the outstanding high temperature properties. Many efforts have been dedicated to this goal for decades [4,5,6,7].

In heat resistant steels, the most important alloying elements are chromium (Cr) for oxidation resistance and nickel (Ni) for strength and ductility. Other elements are added to improve these high temperature properties. Their effects are given generally as follows:

Chromium (Cr) is the one element present in all heat resistant steels. Oxidation resistance comes mostly from the chromium content. Chromium also adds to high temperature strength and carburization resistance. Chromium tends to make the atomic structure "ferritic". Both chromium and iron have the tendency to form ferrite, and which is counteracted by nickel. High chromium also contributes to sigma formation.

Nickel (Ni) added to a mix of iron and chromium increases ductility, high temperature strength, and resistance to both carburization and nitriding. Nickel tends to make the atomic structure "austenitic". Iron base alloys become austenitic when a certain amount of nickel added, so it is the most important element in austenite. Nickel decreases the solubility of both carbon and nitrogen in austenite. The austenitic heat resistant steels content at least 8%Ni.

There are some other metallic alloying elements, such as manganese (Mn), molybdenum (Mo), titanium (Ti), vanadium (V), tungsten (W), aluminum (Al), cobalt (Co), niobium (Nb), zirconium (Zr), Copper (Cu) and the rare earth elements such as boron (B), cerium (Ce), lanthanum (La) and yttrium (Y). These elements added well improve the steels integrative properties at elevated temperature. Some of them for strength, others are largely for oxidation resistance, process workability and microstructure stability.

As to the nonmetallic elements, Carbon (C) is the most important strengthening element, even a few hundredths of a percent. When the carbon level increases, the steel becomes stronger, but it also becomes less ductile. Carbon is controlled within certain limits in heat resistant steels and the limits are related to the processing methods. Most wrought heat resistant steels contain around 0.05 to 0.10% carbon, while cast heat resistant steels usually have from 0.35% up to 0.75% carbon. Carbon dissolves in the alloy and induces solution strength. It is also present as small, hard particles called carbides. These are chemical compounds of carbon with metallic elements, such as chromium, molybdenum, titanium and niobium et al. Nitrogen (N) is a small amount in heat resistant steels, and serves to strengthen both martensitic austenitic alloys. Silicon (Si) decreases the solubility of carbon in the metal (metallurgically it increases the chemical "activity" of carbon in the steels), which is an important variable in the steelmaking process, as a strengthening element, normally above 0.04%. Silicon improves both oxidation and carburization resistance, as well as resistance to absorbing nitrogen for heat resistant steels at high temperature. A silicon oxide layer formed just under the chromium oxide scale on the steels is what helps the alloy resist carburization and nitrogen absorbing. Sulphur (S) is normally regarded as an impurity, and commonly limited below 0.010% in most steels. Sulphur is also detrimental to weldability. But it has the benefit of improving machinability, so it is kept up around 0.02% for 304 and 316 steels. Phosphorus (P) is generally undesirable element in heat resistant steels for brittle effect when it segregates at grain boundary. It is also harmful to nickel alloy weldability. It is normally specified an upper limit for most steels, even as the nickel weld fillers are specified to have no more than 0.015% phosphorus.

2.3 Precipitates

As mentioned above, to great extent, all the heat resistant steels are strengthened by precipitation. Generally, complex alloy steels are more advantageous than those simply alloyed, because they have a predominant characteristic of precipitate reaction. However, the more complex the steel is, the more complicated precipitation reactions.

Most of these precipitates are nitrocarbides, and few of them are inter-metallic compounds, including the intergranular and boundary precipitation as well as variable carbide reactions. The morphology, size and distribution of these precipitating particles are modified by complex alloy elements and which enhance the properties obviously because of carbide reactions accompanied by microstructural and microchemical changes. As to what kinds of precipitates present in a specified steel is bound with its compositions and heat-treatment states.

The kinds of precipitate usually form in ferritic heat-resistant steels are $M_{23}C_6$, M_3C, M_2C, M_6C, MX, Laves and Z-phase. Some extra phases, such as θ, χ and G-phase, may form in austenites. The various precipitates frequently reported in heat resistant steels are given in the following table 1[8,9,10,11].

Precipitate	Structure	Parameter (Å)	Composition
NbC	fcc	a=4.47	NbC
NbN	fcc	a=4.40	NbN
TiC	fcc	a=4.33	TiC
TiN	fcc	a=4.24	TiN
Z-phase	tetragonal	a=3.037 c=7.391	CrNbN
$M_{23}C_6$	fcc	a=10.57-10.68	$Cr_{16}Fe_5Mo_2C$(e.g.)
M_6C	diamond cubic	a=10.62-11.28	$(FeCr)_{21}Mo_3C$; Fe_3Nb_3C; M_5SiC
Sigma	tetragonal	a=8.80 c=4.54	Fe,Ni,Cr,Mo
Laves phase	hexagonal	a=4.73 c=7.72	Fe_2Mo, Fe_2Nb
χ-phase	bcc	a=8.807-8.878	$Fe_{36}Cr_{12}Mo_{10}$
G-phase	fcc	a=11.2	$Ni_{16}Nb_6Si_7$, $Ni_{16}Ti_6Si_7$

Table 1. The main precipitates in heat-resistant steels

There is a simple relationship between carbide structure and the metallic element in the periodic table. It is shown in table 2.

Group / Row	IV	V	VI	VII	VIII		
3	TiC	VC / V_2C	$Cr_{23}C_6$ / Cr_7C_3 / Cr_3C_2	$Mn_{23}C_6$ / Mn_7C_3 / Mn_3C	δ-Fe_2C / Fe_3C / Fe_2C	Co_3C / Co_2C	Ni_3C
4	ZrC	NbC	Nb_2C	Mo_2C / MoC	Cubic	Hexagonal	Orthorhombic
5	HfC	TaC	Ta_2C	W_2C / WC			

Table 2. Relationship between carbide structure and the position of metallic elements.

Table 2 Relationship between carbide structure and the position of metallic elements in the periodic table [12]. At service condition, The most common populations of secondary phases in heat resistant steels are $M_{23}X_6$ carbides, MX carbonitrides, Laves phase and Z phase [13]. So a lot of attention has been paid to these precipitates of $M_{23}C_6$, MX, Laves and Z-phase for their notable effects on creep behavior. Much more details about them are given as follow:
$M_{23}C_6$ is the main precipitates in most heat resistant steels. It is believed that the amount of $M_{23}C_6$ and its size and distribution strong affect the creep strength of creep-resistant steels. Generally this carbide is chromium rich, with Fe, Ni, Mn and Mo substitute for Cr partially, and the fraction of these substitute atoms can be up to forty percent [14]. The crystal structure of $M_{23}C_6$ is a complex face centered cubic (fcc) with a parameter about 1.06nm , which changes slightly for metallic elements variation.
As the main carbides in heat resistant steels, $M_{23}C_6$ precipitates mostly precipitate along grain boundaries during tempering treatment, and some particles might form in the process stage. Even though $M_{23}C_6$ is a very stable precipitation in structure, but they coarsen obviously during creep exposure at elevated temperature. Their average size increases while density decreases with exposure time prolonged. The large particles are growing and the fines dissolving but volume fraction remains constant. Measurement the size distribution and the content change of metallic element were made by researches [15,16] to characterize the microstructure evolution. Investigations [17,18] indicate that B addition has an obvious effect to improve creep strength.
Another typical kind of carbide/nitride nominated as MX is with cubic NaCl-type structure. Its crystal parameter is 0.43±0.017nm for composition variety. Much fine and dispersive MX particles precipitate intergranular and ingranular during tempering and exposure at elevated temperature when the strong forming elements of V, Ti, Nb and W added in alloys. MX carbonitrides usually form on dislocations in the matrix, so they increase creep strength for their dislocation pinning action. Even though MX precipitates are stable and do not coarsens heavily during creep exposure [19,20]. However, it is found that they might dissolve and form complex nitride Z-phase under long-term creep exposure [21].
Z-phase, Cr(V,Nb)N, is another nitride precipitate with similar elements as the MX. Its crystal structure is tetragonal and lattice parameters are a=0.30 nm and c=0.74 nm. In high chromium ferritic steels, a Z phase was found after long-term creep [8]. Additionally, it has

been observed that the Z phase precipitates after creep in both weld metals and the heat-affected zone.

The discovery of Z phase precipitation in a 9–12%Cr steels actually used in power plants raised serious questions about the long term stability. Very little is known about the behavior of Z phase in 9–12%Cr steels, since it has been observed to precipitate only after long times of exposure in plants or after long term creep testing. Z-phase formation has been recently recognized to decrease the long-term creep strength, because the formation of Z-phase consumes fine MX carbonitride particles that are the main strengthening species in 9 – 12% chromium ferritic steels[22]. The Z phase precipitates as large particles, which do not contribute to precipitate strengthening, and thus the creep strength of the steel is considerably lowered.The additions of W and Mo in alloys improve their creep behavior obviously; However, these elements are apt to form intermetallic phase Laves phase $(Fe,Cr)_2(Mo,W)$. Laves phase particles usually precipitate in grain boundaries close to $M_{23}C_6$ carbides in equiaxed shape during creep exposure [23,24]. The crystal structure of Laves phase is hexagonal with a=0.47, b=0.78nm with a small range variety for composition change. It is found that the size of Laves phase particles increase rapidly in higher exposure temperature. There is an argument about the effects of Laves phase on creep property. The precipitation of Laves phase leads to a depletion of these two elements Mo and W in the matrix and to a reduction of the solid solution hardening effect in alloys. On the other hand, Laves phase precipitation can increase the creep strength by precipitation hardening before it coarsens obviously. It believes that the Laves phase has a negative effect when its size becomes as one of the largest precipitates in alloys.

As to what kinds of precipitation present in a specified steel is related to its chemical compositions, heat-treatment parameters. In fact, the heat-resistant steels have shown very complicated metallurgy physics factors, e.g. investigations indicate that the same steel in different heats or the same steel exposed in different conditions and experiences may contain different carbides with dissimilar precipitation, morphology or distribution. That is why the metallurgists face a challenge when they try to evaluate the residual life from the appearance of precipitation and microstructure evolution.

3. Classes of heat-resistant steels and applications

Heat-resistant steels have chemical stability, enough strength and gas corrosion-resistance. They can be divided into low alloy steels, martensitic steels, and austenitic steels according to their chemical composition and microstructure.

3.1 Low alloy steels

Due to good mechanical properties at elevated temperatures and sufficient corrosion resistance, low alloy steels are widely used in pressure part applications in boilers for thermal power plants. This type steels are especially used in thick-section components such as headers and steam pipes. Particularly these are tubing steels for the cooler sections of superheaters and reheaters and also for the waterwalls.

Traditionally, in power generation industry, this class alloy steels Grades 11 and 22 have been used for waterwalls, superheaters and thick-section parts in boilers; and 1CrMoV for rotors for many years[25,26].grade 22 (2.25Cr1Mo) developed in 1960' and applied in worldwide power plants for large amount of components: tubes, pipes, cast, forged.

With the increased steam pressures and temperatures, the rising of operating parameter in power plants also have an impact on the conditions for operating water walls. Therefore, the former standard materials used for water walls like T/P12 (13CrMo4-4) do no longer meet the requirements of advanced boilers. Since steels used in water wall panel should allow welding without post weld heat treatment (PWHT), the water wall panel is too large to perform a heat treatment. This demand also excludes high-strength steels martensitic steels. Two new steel grades present and satisfy all requirements. They are T/P23 originated in Japan and T/P24 (7CrMoVTiB10-10) developed by Vallourec and Mannesmann (V&M) in Germany. W. Bendick et al [27] pointed out that the characteristics of both steel grades have potential application.

In the years '80-90 many efforts have been devoted to increase the plants performances and therefore materials with enhanced performances were requested. The metallurgists and the material researchers started to develop new chemical compositions on the base of existing steels with the addition of elements (V, W, Nb, Ti) able to give strengthening by precipitation [3]. The new advanced low alloy Grades 23 and 24 steels, improved based on Grades 22, are strong candidates for new power and petrochemical plant construction and for the eventual large-scale replacement of steam pipework on existing power plant. Grade 23 is a modified 2Cr1Mo steel and it is claimed that does not require post-weld heat treatment. The improvements have been achieved through the addition of tungsten and boron whilst reducing the carbon content. A further recent development has been Grade 24, where the microstructure has been further refined by modification of the vanadium, niobium, titanium, boron, tungsten and molybdenum levels. This steel has superior long-term creep properties to Grade 23 at 500 °C; is equivalent to Grade 23 at 550 °C, and is intermediate between Grade 22 and 23 at 600 °C.

The main differences in compositions are described in the following table 3 based on ASTM A213 Standard.

Grade		C	Mn	P	S	Si	Cr	Mo	W	Nb	V	B	Other
22	min	0.05	0.30	-	-	-	1.90	0.87	-	-	-	-	-
	max	0.15	0.60	0.025	0.025	0.50	2.60	1.13	-	-	-	-	
23	min	0.04	0.10	-	-	-	1.90	0.05	1.45	0.02	0.20	0.0005	N: 0.03
	max	0.10	0.60	0.030	0.010	0.50	2.60	0.30	1.75	0.08	0.30	0.0060	max
24	min	0.05	0.30	-	-	0.15	2.20	0.70	-	-	0.20	0.0015	N: 0.012 max
	max	0.10	0.70	0.020	0.010	0.45	2.60	1.10	-	-	0.30	0.0070	Ti: 0.06-0.10

Table 3. The compositions of low alloy steels based on ASTM Standard.

The effect of adding or changing the level of different elements on the microstructure and properties of the alloys is obvious. Due to the same basic composition and the same structure T/P23 and T/P24 mostly have similar physical properties (including the oxidation resistance) as T/P22. This is true for the physical properties. However, large differences exist for the mechanical properties including creep. The alloying additions have led to a strong increase of strength. The values of minimum yield strength for T/P23 and T/P24 are higher by almost a factor of 2 compared to T/P22. The most important property of a heat resistant steel is its creep rupture strength. Similar to the tensile properties the steels T/P23 and T/P24 reveal considerably higher values. Test results at 550°C are shown in figure 3. The rupture points are close to the mean lines of the steel grades. Figure 4 shows the 10⁵h creep rupture strength of T/P24 is only slightly below T/P91 in the lower temperature range. Although the values for T/P23 are somewhat lower at 500–550°C, they still lie considerably above T/P22. The T/P23 curve approaches T/P24 with increasing temperature and crosses it at 575°C. Due to the limited oxidation resistance of the two steels, it is not recommended to use them at temperatures higher than 575°C for long-term service.

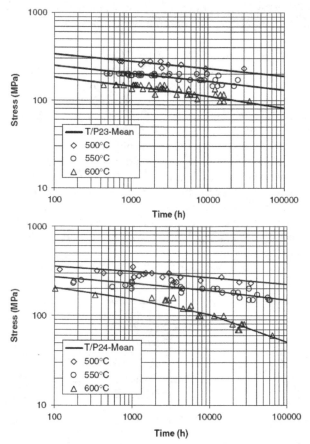

Fig. 1. Creep test on Grades 23 and 24 for the temperature range 500-600 °C [from ref.31]

Alloy	Mechanical properties			Alternative nomenclature
	Yield Stress, MPa	UTS, MPa	Elongation, %	
Grade 11	205	415	30	P11/T11/13CrMo 4 4
Grade 22	205	415	30	P22/T22/10CrMo 9 10
Grade 23	400	510	20	P23/T23/HCM2S
Grade 24	580	670	20	T24/7CrMoVTiB 10-10

Table 4. Mechanical properties for low alloy steels at room temperature

The most recent advancement in low alloy steel is the development of 3Cr–3W(Mo)V steels, which has a higher creep strength than 2.25Cr–1Mo steel (T22) and 2.25Cr–1.6W–VNb steel (T23) [28,29]. Fig.2 shows the extrapolated 10^5 h creep rupture strength for T22, T23, T24 and 3Cr–1.5W–0.75Mo–0.25V without Ta (Grade A) and with 0.1Ta (Grade B) as a function of temperature, comparing with martensitic T91 steel. The creep rupture strength of the Grade B steel is higher than T23 and T24 steels for the entire test temperature range and also higher than T91 steel up to 615°C.

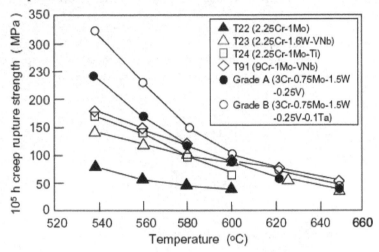

Fig. 2. Creep rupture strength of 10^5 h as a function of temperature of Grade A and B of 3Cr–1.5W–0.75Mo–0.25V steel, comparing with T22, T23, T24 and T91 [from ref.6].

In general, Cr–Mo low-alloy ferritic steels are tough and ductile at lower operating temperatures and maintain good strength at higher temperatures. Unfortunately, when subjected to prolonged exposure to intermediate service temperatures, these steels can become embrittled with an associated decrease in fracture toughness and a shift in ductile-to-brittle transition temperature (DBTT) to higher temperatures [30,31]. The embrittlement is mainly caused by changes in the microchemistry of grain boundaries, which is referred to as temper embrittlement. Temper embrittlement is non-hardening embrittlement and arises from grain boundary segregation of impurity elements such as P, Sb and Sn as a result of long-term exposure in the temperature range of 350–600°C. Several reports reveal that phosphorus is the major one of these embrittling impurity elements in steel.

Another type of low alloy extensively used for various engineering components in thermal power plants of China, India, Russia and former Soviet Union countries and Eastern European countries Cr1Mo steels, such as 12Cr1MoV, 12X1MΦ(Russian Grade), 14CrMo4-5 (ISO 9328-2, 1991), 13CrMo4-5 (EN 10028-2, 1992), or 12C1.1 (ASTM A182-96) et al. They are the heat resistant steels with lowest of alloy additions in chemical composition. Generally, the pipelines used to transport superheated steam in the temperature range 500 – 560 °C and under a pressure, P = 10 – 15 MPa.

The initial microstructure of power plant low alloy steels is ferrite bainite or ferritie-pearlite [32,33]. Generally, The Cr–Mo and Cr–W elevated-temperature steels are used in the normalized-and-tempered condition. Normalizing consists of austenitizing by annealing above A1, the equilibrium temperature where ferrite (body-centered-cubic structure) transforms to austenite (face-centered-cubic structure), and then it is air cooled. In low alloy steels with <5%Cr, bainite (ferrite containing a high dislocation density and carbides), polygonal ferrite, or a combination of these two constituents form, depending on the section size of produces. Their creep strength enhanced by the formation of precipitates, which are stable alloy carbides and intermetallic compounds obtained following normalizing heat treatment later on subjected to very severe tempering (about 700°C for several hours).

Steels destined for power plant applications might contain any of the following precipitates: carbides or carbonitrides M_3C , $M(C, N)$, $M_2(C, N)$, M_7C_3, $M_{23}C_6$, M_6C, Laves phase (M stands for metallic solute atom) and intermetallic precipitates. It was determined in that the precipitation sequences at elevated temperature for steels are as follows [34,35]:

a.   steel 2.25Cr1Mo

$$M_3C \rightarrow M_3C + M_2C \rightarrow M_3C + M_2C + M_7C_3 \rightarrow M_3C + M_2C + M_7C_3 + M_{23}C_6;$$

b.   steel 3Cr1.5Mo

$$M_3C \rightarrow M_3C + M_7C_3 \rightarrow M_3C + M_7C_3 + M_2C + M_{23}C_6 \rightarrow M_7C_3 + M_2C + M_{23}C_6.$$

c.   steel Cr1MoV

$$M_3C + MC \rightarrow M_3C + MC + M_{23}C_6 \rightarrow M_3C + MC + M_{23}C_6 + M_7C_3 \rightarrow MC + M_{23}C_6 + M_7C_3.$$

The precipitation sequence in different low alloy steels is obviously different in the evolution of carbide precipitation even the thermodynamic driving forces are apparently similar. It is believed that the precipitation sequence difference is related to the changeable driving force for various precipitates in different steels.

During long time service in creep regime to such conditions the microstructure of steel changes, bainite/pearlite decomposes as well as carbides precipitation at the grain boundaries and carbides coarsening processes proceed. Structure changes cause formation of cavities and development of internal damages [36]. It is well known that there is a close coherence between changes in microstructure and deterioration of mechanical properties. So many attentions have been paid to investigation of the carbides precipitation kinetics of power plant heat resistant steels during ageing or long-term service at elevated temperatures. The aim on purpose is try to determine any microstructural parameters that may be used to estimate service history and may be practicable for assessment of remnant life of equipments.

There is a demand for higher strength steels for waterwalls to improve the creep behavior at the higher temperatures in supercritical plant. With the continuing effort to achieve higher efficiencies and lower costs in power generating components, the materials of Grade 23 and 24 and any alloy with higher strength but with no requirement for post-weld heat treatment would be an attractive material. It would be significant improvement.

3.2 Martensitic/ferric steels
Martensitic heat-resistant steels include medium-chrome steel containing 5-9% Cr and high-chrome steel containing 12% chrome, which have been introduced into power plant

materials more than half century. The 9-12% Cr martensitic steels are currently used in both boilers and in steam turbines for many components. In boilers, these steels are used for tubing in superheaters and reheaters, operating with metal temperatures up to above 620°C currently. The thick-section parts such as headers and steam pipes are also fabricated from these materials. It has been recognized that the 9-12% Cr steels are the key materials to increase the thermal efficiency of steam power plants. In the last three decades, a number of new 9–12%Cr steels with improved creep strength have been developed for long-term service at temperatures close to 650°Cin high-temperature components of ultra supercritical power plants [37]. This is a great progress to develop high creep strength and corrosion resistance steels at ever increasing temperature. The steadily improved creep rupture strength of new martensitic 9–12% Cr steels has been used to construct new advanced fossil-fired steam power plants with higher efficiency. The applications of these new alloys achieve not only high efficiencies, but also reducing the emission of CO_2 and other environmentally hazardous gases at least 20% [18,38].

Increase in steam parameters from subcritical 180 bar/530–540°C to ultra-supercritical values of 300 bar and 600°C has been realised, and this has led to efficiency increases from 30–35% to 42–47%, equivalent to approximately 30% reduction in specific CO_2 emission.

For the long-term application of the new steels, it is necessary to assess the microstructural changes that are likely to occur during service exposure and to evaluate the effect of such changes on the hightemperature creep behavior. Only with this information can the design values for components be correctly assigned.

In general these alloys have lower coefficients of thermal expansion and higher thermal conductivities than austenitic steels and should therefore be more resistant to thermal cycling.

The development of 9–12% chromium steels is originated a century ago with the manufacture of a 12% Cr and 2–5% Mo steel for steam turbine blades by Krupp and Mannesmann in Germany [39]. The high-chromium, high-carbon martensitic steels were hard and were subsequently developed commercially for applications such as cutlery knives, razors, scalpel blades, and heat-resisting tools and bearings in competition with austenitic stainless steels.

Beginning with the 12CrMoV steel introduced in power plants in the middle of 1960s, steel development over the past decades has led to new steam pipe steels like the modified 9Cr steel P91, introduced in plants in 1988s, to the tungsten-modified 9Cr steels P92 introduced in plants in 2001 and E911 introduced in 2002. Similar steels have been developed and applied for large forgings and castings of steam turbines. The 9 -12% Cr steels with lower carbon (0.1% max) contents and additions of Mo, W, V, Nb, N and other elements, possessing higher creep-rupture strengths combined with good oxidation and corrosion resistance at elevated temperatures, have subsequently been developed. These steels have been used or considered for use in electrical power plants, petrochemical and chemical plants, gas turbine engineering, aircraft and aerospace industries, and as nuclear fission and fusion reactor components as well.

In order to develop these new alloys with advanced characteristics, many efforts have been made to investigate the metallurgy mechanism. Many strengthening mechanisms have been proposed to explain the improved creep strength. Much more attentions have been paid to research the effects of various additions, for instance, in the course of development of 12 CrMoV steels nickel was added to improve impact properties and to suppress the presence of δ-ferrite in the microstructure. It also was attributed to solid solution hardening and a reduced solubility of carbon. However, the excessive amounts of nickel, greater than 0.6 wt. %, caused an accelerated reduction in the creep rupture strength, which maybe partly attributed to reduced stability of M_2X phase and precipitation of M_6X particles [40]. Mod.9Cr–1Mo steels have attained their high creep strength by the addition of V, Nb and N which form fine precipitates MX, and the creep strength of high Cr ferritic steels have been improved further by replacing part of Mo with W[41,42,43].

Specific alloys in this class generally include 12Cr and 9Cr martensitic steels. One famous 12Cr steel is X20CrMoV12.1. Since the X20CrMoV12.1 was developed in 1950s, it has been successfully used in power plants over several decades up to temperature of about 566°C. The creep strength of X20 is based on solid solution strength hardening and precipitation of $M_{23}C_6$ and MX carbides. Specific alloys in 9Cr class include Grade 9, Grade 91, E911, Grade 92, HT 9, HCM12 etc and the compositions are given in following Table 5.

Steel Type	Desig-nation	Country of Origin	Analysis(wt. %)										
			C	Si	Mn	Cr	Ni	Mo	V	Nb	W	N	B
9Cr-1Mo	9Cr-1Mo*	UK	0.10	0.70	0.50	9.5	0.20	1.0					
	T9	Japan	≤0.15	0.25-1.00	0.30-0.60	8.0-10.0		0.90-1.10					
	EM10*	France	0.10	0.30	0.50	9.0	0.20	1.0					
9Cr-2Mo	HCM9M*	Japan	0.07	0.30	0.45	9.0		2.0					
	NSCR9*	Japan	0.08	0.25	0.50	9.0	0.10	1.6	0.15	0.05		0.030	0.003
	EM12	Belgium/France	0.08-0.12	0.30-0.50	0.90-1.20	9.0-10.0		1.9-2.1	0.25-0.35	0.35-0.45			
	JFMS*	Japan	0.05	0.67	0.58	9.6	0.94	2.3	0.12	0.06			
9Cr-MoVNb	Tempaloy F-9	Japan	0.04-0.08	0.25-1.00	0.40-0.80	8.0-9.5		0.90-1.1	0.15-0.45	0.20-0.60			≤0.005
	T91	USA	0.08-0.12	0.20-0.50	0.30-0.60	8.0-9.5	≤0.40	0.85-1.05	0.18-0.25	0.06-0.10		0.030-0.070	
9Cr-MoVNbW	COST 'B'*	Europe	0.17	0.10	0.10	9.5	0.10	1.5	0.25	0.05		0.005	0.010
	E911*	Europe	0.10	0.20	0.40	9.0	0.20	1.0	0.20	0.08	1.0	0.070	
	TF1*	Japan	0.12	0.20	0.50	9.0	0.80	0.60	0.26	0.06	1.6	0.050	0.003
	TB9(NF616) (T92)	Japan	0.07-0.13	≤0.05	0.30-0.60	8.5-9.5	≤0.40	0.30-0.60	0.15-0.25	0.04-0.09	1.5-2.0	0.050-0.070	0.001-0.006

Table 5. Compositions for martensitic steels

Alloy development programs have concentrated on improving the creep properties of 9-12Cr steels for decades. Based on the Grade 9, the aim to improve creep resistance has been resulted in alloys such as Grade 91,92, E911, Grade 122 and TAF Fig.3 shows the extrapolated 105 h creep rupture strength for some 9-12Cr martensites as a function of temperature. The creep rupture properties of P92 and TAF have been improved obviously.

An outstanding position is held by TAF steel with relatively high Cr-contents of 10.5 wt% and B-contents of 0.027– 0.040 wt%. However, the other attempts to develop high strength alloys of this type to promote creep strength maybe achieved, but these developments fail due to a lack of long-term stability at much higher temperature. It is believed that steam oxidation resistance might limit the maximum operating temperature of these materials, so alloy development is now focused on developing materials with higher chromium contents with a creep strength equivalent to Grade 92.

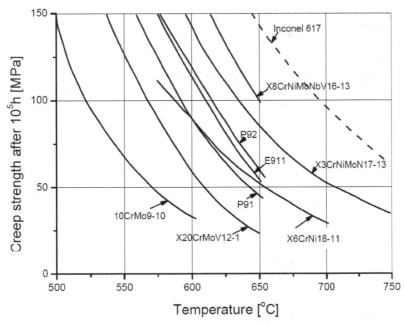

Fig. 3. Creep Strength of Exposed (10^5h) Materials as a Function of Temperature [ref.44].

Many investigations have shown that alloying additions appear to be the effective way to improve the creep strength of the 9–12%Cr steels. Fujita[45] found a substantial improvement of creep strength by adding B, and he confirmed the positive effect of 0.040 wt.% B in creep of TAF steel up to 130,000 h at 650 °C, where the target of 10^5 h creep life at 650°C and 100MPa was almost achieved [46]. Apart from B, Cr is a key element influencing both the oxidation resistance and the creep strength. Experience shows that 12 wt% Cr are needed to raise the oxidation resistance to an expected level [47].

Work carried out in Japanese and European programmes has resulted in the improved 9%- and 12%-Cr alloys that can be used for thick-section components intended for use in the operating range 565-620°C. Materials such as Grade 92 and Grade 122 are used in sub and supercritical units where their higher strengths allow the use of thinner sections, thereby reducing the threat of fatigue due to thermal cycling. The reduction in weight brought about by thinner walls also has the effect of reducing stresses at the boiler and turbine connections, as well as on the structural steel work, all of which contributes to increased life of components and reduced costs. The European companies and research institutes carried out a joint program (COST 501) to develop and qualify modified 10%Cr steels. Results showed that the fatigue properties of the 10%Cr material were improved compared to conventional 11-12 %Cr steels. An added advantage of the 10%Cr material compared with conventional low alloy steels is that the lower thermal-expansion coefficient allows a greater temperature rise to be tolerated at start-up. Another aim of the new COST 522 programme is to develop 9-12%Cr steels for applications up to 650°C.

An extensive study [48] of the microstructural development of these alloys during thermal exposures has shown a starting microstructure of elongated dislocation cells and sub-grains aligned with $M_{23}C_6$ particles, together with smaller VN and M_2X particles inside the sub-grains.During long-duration creep tests some softening of the material occurred in the specimen threads due to thermal exposure. Some of this softening was associated with Oswald-ripening of existing particles but precipitation of new particles, in particular Laves phase, was observed. In parallel, the dislocation density decreased such that few dislocations were observed inside the sub-grains. All these microstructural changes occur more slowly in steels that contain boron [49].

Alloy	Composition, wt%									
	C	Mn	Si	Cr	Mo	V	N	W	B	Other
Grade 9	0.12	0.45	0.6	9	1					
E911	0.12	0.51	0.2	9	0.94	0.2	0.06	0.9		0.25Ni
Tempaloy F12M				12	0.7			0.7		
Grade 91	0.1	0.45	0.4	9	1	0.2	0.049			0.8Ni
Grade 92	0.07	0.45	0.06	9	0.5	0.2	0.06	1.8	0.004	
HCM12	0.1	0.55	0.3	12	1	0.25	0.03	1.0		
Grade 122	0.11	0.6	0.1	12	0.4	0.2	0.06	2.0	0.003	1.0Cu
TAF	0.18			10.5	1.5	0.2	0.1		0.04	0.05Ni
TB12	0.08	0.5	0.05	12	0.5	0.2	0.05	1.8	0.3	0.1Ni
NF12	0.08	0.5	0.2	11	0.2	0.2	0.05	2.6	0.004	2.5Cu
SAVE12	0.1	0.2	0.25	10		0.2	0.05	3.0		3.0Co, 0.1 Nd
X20CrMoV121	0.20	1.0	0.5	12	1.0	0.3				0.6Ni
X12CrMoVNbN101	0.12			10	1.5	0.2	0.05			
X18CrMoVNbB91	0.18			9	1.5	0.25	0.02		0.01	

Table 6. Compositions of martensitic steels

The microstructure for these 9-12Cr materials is tempered martensite with creep resistance imparted by controlled precipitation of carbides and nitrides. In general, 9-12Cr steels are also used in the normalized-and-tempered condition. Fig.5 shows a typical TEM micrograph of tempered 9Cr martensite. Martensite forms when products are normalizing by austenitizing above A1 and then air cooled. During this process, the formation of δ-ferrite should be avoided as this may cause embrittlement, resulting in fabrication problems. Subsequent tempering above 700°C for hours, the strengthening precipitates form. The types of precipitates formed in martensitic heat-resistant steels also depend on the composition, temperature history during fabrication, and time and temperature of service exposure. The main precipitate in the 9–12%Cr steels is the $M_{23}C_6$ carbide consisting of Cr, Fe, Mo, (W), and C. This carbide produces the basic creep strength of 9-12%Cr steels by precipitating on subgrain boundaries during tempering. The $M_{23}C_6$ carbides increase creep strength by retarding subgrain growth, which is a major source of creep strain in 12%Cr steels. The thermal stability of $M_{23}C_6$ is relatively high. The MX precipitates in 12%Cr steels consist mainly of V, Nb, and N, which precipitate within subgrains where they pin down free dislocations and increase creep strength. Thermal stability of the MX precipitates is very high, leading to high creep strength. It is interesting to note that equilibrium calculations indicate that the strengthening effect of vanadium in matensites is by precipitation of vanadium nitrides. This means that even though nitrogen is not a specified alloying element in these steels, the tramped nitrogen contributes to the high strength.

Their compositions and mechanical properties are given in Table 6 and table 7.

Alloy	Mechanical properties			Alternative nomenclature
	Yield Stress MPa	UTS, MPa	Elongation %	
GRADE 9	205	415	30	P9/T9/STBA26
E911				
Tempaloy F12M	470	685	18	
GRADE 91	415	585	20	X10CrMoVNb 91, GRADE 91
GRADE 92	440	620	20	P92, T92
HCM12				SUS410J2TB
Alloy 122	400	620	20	HCM12A
TAF				
TB12				
NF12				
SAVE12				
X20CrMoV121	495	680	16	
X12CrMoVNbN101				
X18CrMoVNbB91				

Table 7. General mechanical properties for typical martensites

Fig. 4. TEM micrograph of tempered 9Cr martensite

3.3 Austenitic steels and other materials

Austenitic stainless steels are FeCrNi alloys with chromium content more than 13wt% with an austenitic structure at room temperature. Austenitic steels are more expensive than ferritic steels for their high alloy element additions. Traditionally, applications for austenitic steels are restricted to the higher temperature boiler tubes as well as to specific situations where severe corrosion conditions occur. Austenitic materials are often used as weld overlay on ferritic materials to repair corroded areas or to provide protection in areas where corrosion could be a problem.

Austenitic steels are developed based on 18%Cr-8%Ni, originated from AISI 302. In order to obtain the required properties, other alloying elements are added. Except the interstitial elements such as carbon and nitrogen, some substitutional elements such as Mn, Mo, W, Cu, Al, Ti, Nb, V, etc are added. These alloying elements are classified as ferrite stabilizers or austenite stabilizers by their effects to promote a ferritic structure or an austenitic structure. Their contributions can be evaluated using the notation of chromium and nickel equivalents as presented below [50]:

$$Ni_{eq}=Ni+Co+0.5Mn+0.3Cu+30C+23N(mass\%)$$

$$Cr_{eq}=Cr+2.0Si+1.5Mo+1.75Nb+1.5Ti+5.5Al+0.75W(mass\%)$$

Various types steels of this class have been produced by increasing the strength with the addition of alloying elements. Table 8 and table 9 give the composition and properties of common austenitic steels. Nb, Ti and V can greatly improve the creep strength of austenitic stainless steels by precipitating fine carbides or carbonitrides. Furthermore, addition of these elements can stabilize the alloy against intergranular corrosion. Metallurgical physics of austenite development is to increase the volume fraction of strengthening precipitates by replacing chromium carbides with other more stable carbides, which, at the same time, frees chromium back to the matrix to give improved corrosion resistance. However,

Alloy	Composition, wt%											Other
	C	Mn max	P max	S max	Si max	Cr	Mo	N	Nb	Ni	B	
AISI 302	0.15	2	0.045	0.03	1	18				9		
AISI 304	0.08	2	0.045	0.03	1	18				8		
AISI 321	0.08	2	0.045	0.03	1	17				11		0.15Ti
AISI 347	0.08	2	0.045	0.03	1	17				11		0.8 Nb+Ta
AISI 316	0.08	2	0.045	0.03	1	17	2.5			12		
AISI 309	0.2	2	0.045	0.03	1	23				14		
AISI 310	0.25	2	0.045	0.03	1.5	24				19		
ASME TP347HFG	0.08	1.6			0.6	18			0.8	10		
Tempaloy A-1	0.12	1.6			0.6	18			0.1	10		0.08 Ti
Tempaloy A-3	0.05	1.5			0.4	22		0.15	0.7	15	0.002	
Super304H	0.1	0.8			0.2	18		0.1	0.4	9		3.0 Cu
HR3C	0.06	1.2			0.4	25		0.2	0.45	20		
Hr6W	0.10	2.0	0.03	0.03	1.0	23			0.4 max	40		6.0 W, 0.2 maxTi
NF709	0.15	1			0.5	20	1.5		0.2	25		0.1 Ti
Esshete 1250	0.09	6	0.03	0.003	0.6	15	1	0.04		10	0.004	0.02Al, 0.9Nb, 0.25V, 0.14Cu

Table 8. Chemical composition for different austenitic steels

investigations have reported that these the presence of secondary phase particles can strengthen austenitic steels, but these secondary phase particles has also been seen to affect the fatigue behavior.

Table 8 shows the compositions of some heat resistant austenitic stainless steels, mainly AISI 300 series alloys. Austenitic steels such as AISI 316 and 304 are used extensively as structural materials in heavy sections for pressure vessels and pipes in power plant. In fact, austenitic steels are used in many areas, which are subjected to varying temperatures and temperature gradients. Austenitic steels generally have low thermal conductivities and high coefficients of thermal expansion, it is noticed that the high thermal stresses can develop resulting in fatigue cracking.

Alloy 800HT is an austenitic nickel–iron–chromium alloy. This alloy is characterized by high creep strength and very good resistance to oxidation. Super austenitic stainless steels 253MA or UNS S30815 and 353MA or UNS S35315 are austenitic chromium–nickel steels alloyed with nitrogen and rare earth metals. They have high creep strength and very good resistance to isothermal and, above all, cyclic oxidation. Fig.5 presents the creep strength of some auatenitic steels at temperature above 600°C.

In the recent decades, the developing high strength ferritic steels have been used instead of the austenitic steels for their cost-effective, good weldability and fracture toughness, but there is still a place for these austenitic steels which are primarily used in the place where oxidation resistance and fireside creep become more important. Especially with the global increase in energy consumption requires more energy production. Meanwhile the concern on the environmental impact from energy production is continuously focusing. Today, coal-fired thermal power generation is still the most important methods in the foreseeing long-term future, as coal is available at a competitive price and satisfy to safety way. However, the biggest challenge facing coal-fired power plants is to improve their energy efficiency. This can be accomplished by increasing the maximum steam temperature and the steam pressure. Conventionally, the heat efficiency of coal-fired power plants has stayed at around 41% in the super critical (SC) condition with a temperature of 550°C and pressure of 24.1 MPa. In order to attain a power generating efficiency of about 43%, ultra super critical (USC) conditions with a steam temperature at about 600°C should be reached. By increasing the temperature from 550 to 600°C (at most present power plants) to 650–700°C (at next generation power plants), the power plant efficiency can be increased from 36% to more than 50% and the CO_2 emission can be reduced about 30% [51]. Currently, in a power plant operating at SC condition and lower, the ferritic/martensitic heat resistant steels are the dominant materials for steam generation and partial for boiler. In USC operating conditions, the ferritic/martensitic materials (including the higher chrome steels) do not have sufficient creep rupture strength and resistance to high temperature corrosion. Austenitic stainless steels are therefore used as the dominant materials. Although nickel base alloys can meet the requirement, they are too expensive to application.

Alloy	Mechanical properties			Alternative nomenclature
	Yield stress MPa	UTS MPa	Elongation %	
AISI 302	240	585	60	
AISI 304	215	505	70	
AISI 321	240	585	55	
AISI 347	240	620	50	
AISI 316	250	565	55	
AISI 309	275	655	45	
AISI 310	275	655	45	
ASME TP347HFG	205	550	35	
NF709	270	640	30	SUS310J2TB
Esshete 1250	272	359	45	

Table 9. Room temperature mechanical properties for austenitic steels.

The materials used for the next generation power plants are expected to have even higher yield strength at elevated temperature, creep strength (typically 100,000 h rupture strength of around 100 MPa) and high temperature corrosion resistance [52]. However, no current heat resistant alloy can meet these requirements. Therefore, the EU-project 'Advanced (700°C) PF Power Plant – AD700' to develop new advanced heat resistant materials for next generation power plants has been organised. One of the aims was to develop materials for superheaters and reheaters in USC boilers for use at temperatures up to 700°C. One material successfully developed is the austenitic stainless steel grade UNS S31035 (Sanicro 25) [53]. This material provides very high creep strength and good corrosion resistance at high temperatures. Another material, Alloy 800HT is a candidate material for generation IV nuclear power plants.

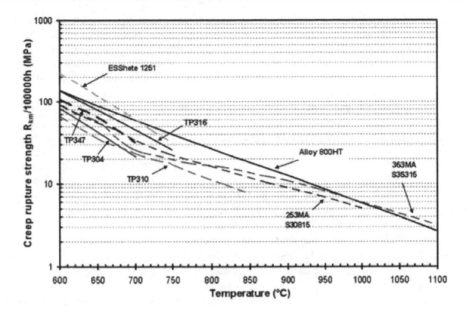

Fig. 5. Creep strength of some auatenitic steels at 100,000 hours [54].

A major concern that remains with the use of austenitic steels is how to join the materials to components manufactured from other material classes. Publishes [55,56] discussed the welding technique and process as well as the behaviour of transition joints between austenitic and martensitic/ferritc steels. They found that the location of the failure varied with test parameter. In the high stress at relative low temperature regime failure usually occurred in the parent ferritic material or weld metal, and failure was close to the weld interface (heat affeced zone) when low stresses were applied over a range of temperatures.

The relatively high costs of these materials coupled with the disadvantages of high thermal expansion coefficient and poor thermal conductivity, will probably continue to limit their applications in advanced power plant. There are two possibilities for further development.Firstly, there could be some value in developing a low-cost austenitic steel, possibly based on Mn rather than Ni. Secondly, the modified austenitic materials developed at ORNL 33-34 could be introduced commercially. The high strength capability of these alloys could give advantages in terms of thinner sections and improved heat-transfer characteristics.

As to other heat resistant materials maybe used in power plant in the future, Oxide Dispersed Strengthened (ODS) Alloys are promised, which are characterised by good creep strength at high temperatures (i.e.>1000°C). Thus the alloys are seen as being useful for higher temperature applications where the strength of the superalloys was inadequate. These alloys are commonly produced by mechanical alloying. Because the processing route was complex and difficult to control in large-scale processing and as a result these alloys were characterised by poor reproducibility of microstructure and inconsistency of mechanical properties. ODS alloys can be classed in iron base and Ni base for the difference of main composition. Researches indicates their good high temperature behavior oxidation resistant is correlated with the dispersive oxide, such as Y_2O_3

It is believed that ODS materials will be most beneficial in areas such as combustor cans and heat exchanger tubing where high temperatures are commonly found but the loading is low incomparison to other areas. Efforts to improve strength and to develop better fabrication techniques will continue, and other major effort will be on the establishment of reliable joining techniques. In addition, it will address advanced joining methods, including laser welding, brazing and coating systems. More attention to improve oxidation resistance will also continue, especially paid to Fe-base materials.

4. Microstructure evolution in inservice materials

The creep properties of heat-resistant steels are controlled by chemical composition and microstructure of these steels. If the chemical composition is given, the microstructure of these steels depends on the heat treatment, temperature and time of creep exposure [57]. The most important strengthening mechanisms in these steels, operating during high temperature creep exposure, are precipitation strengthening and solid solution strengthening. Precipitation strengthening in ferritic steels is predominantly affected by the dispersive MX particles. It has been shown that both the proof stress at room temperature and creep rupture strength increase while density of second phases increase. At the same time, the creep rate decreases.

It is clear that the mechanical properties of heat-resistant steels would deteriorate when it exposed at elevated temperature for long-term. Correspondingly, their microstructure also degrades obviously.

As to the characteristic of microstructure evolution, the main phenomena of degradation are given below:
1. Precipitates coarsening and phase transformation;
2. The original microstructure decomposition;
3. Microviods forming at grain boundaries.

These microstructural evolutions directly connect the deterioration of creep strength and other properties, so they are often considered as a demonstration of overheating exposure, and have been commonly accepted as a qualitative thermal degradation index.

From the investigations in past decades, different class material has different characteristic in microstructure evolution.

4.1 Microstructure evolution in ferritic steels

The low alloy steels or pearlite/bainite steels show the tendency to pearlite/bainite spheroidisation after long-term exposure at high temperature. The typical lamellar structure in CrMoV pearlite change to particle structure. Service exposure has a considerable effect on strength and ductility. Classification of microstructural deterioration in steels has also been established and adopted as regulations, such as carbides spheroidization in 12Cr1MoV steel, Fig.6 gives the states of carbide spheroidization in 12Cr1MoV steel according to the spheroidization evaluation standard [58]. Five levels of degradation are assigned based on the development of carbide spheroidization. The classification specifies Level 1 as having no spheroidization, Level 2 as having slight spheroidization, Level 3 as having medium spheroidization, Level 4 as having complete spheroidization and Level 5 as having serious spheroidization. From Level 1 to Level 5, the lamellar structure in pearlite changes to a particle structure. The pearlite structure disappears at Level 4 and Level 5, which results in deterioration of mechanical properties.

Fig. 7 gives an example of the microstructure in a boiler header in service for 25years. It clearly shows that the degree of spheroidisation in final superheater outlet is much visible than that in the primary superheater outlet. The operating temperature in the former is

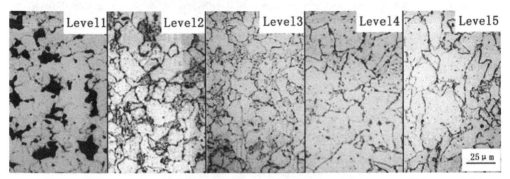

Fig. 6. The evolution standard of carbide spheroidization in 12Cr1MoV steel [from ref.58].

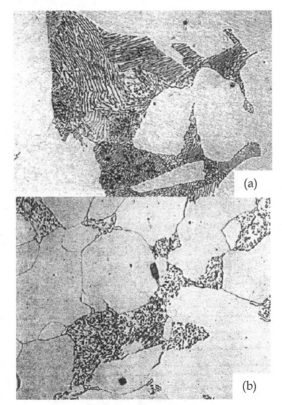

Fig. 7. Microstructure of carbon steel boiler headers in different portions that have been in service for over 25 years (a) primary superheater outlet; (b)final superheater outlet.[from ref.59]

about 30 degree higher than the latter one might be the reason for the difference of microstructure evolution. So the classification of microstructural deterioration in low alloy steels was established by the spheroidization grade. An Italian new reference document for life extension of creep exposed components [60] introduces a classification defined as microstructure evolution based on a reference example relevant to low-alloy steel. The classification with 6 different levels is reported in figure 8. It is easy to judge the header degradation in Fig.7 is about stage B for primary superheater outlet and stages C for final superheater outlet.

Level A - Ferrite and lamellar pearlite

Level B - Initial spheroidisation and carbides precipitation at grain boundary

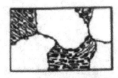

Level C - Intermediate level of spheroidisation, pearlite is partially globular but lamellar structure is still evident

Level D - Pearlite completely spheroidised but carbides still included in primary pearlitic grains

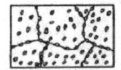

Level E - Carbides homogeneously distributed (no evidence of ferrite and pearlite original structure)

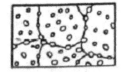

Level F - Carbides homogeneously distributed and partially coarsened through coalescence

Fig. 8. ISPESL classification for microstructure evolution of ferritic steels

Many studies have been conducted on the evolution of carbides present in steels due to creep exposure [14,61,62,63,64]. Separation and coarsening of carbides are considered as an index of material degradation due to creep exposure.

Microstructure evolution in 9-12Cr martensitic steels shows much complicated characteristics during long-term exposure at elevated temperature. From reports in past decades, much more attention was paid to precipitates coarsening, martensitic substructure transformation and microvoids formation at the stage when material degradated seriously [65,66]. Many articles present estimation of average particle size and remaining life fraction for ferritic steel. In particular the attention has been oriented to $M_{23}C_6$, VC, Laves phase and their compositions. However, different results of particle size and composition have been obtained from different author in terms of absolute size value. Statistical measures have a good agreement in terms of correlation of average particle size on temperature exposure or on time maintaining at a certain temperature. This significant difference in terms of absolute values (size, content, interparticle space etc) is not only strongly influenced by statistical approach in estimating average quality, accuracy in measurement at high magnification and sampling technique adopted, the exposure history of the examined component and its metallurgy factors and forming process are essential causations.

Fig. 9. TEM micrograph of exposed X20CrMoV12.1 for 165,000 hours at 550°C [from ref.67].

Unlike low alloy steels, Martensites infrequently is less sensitive to form intergranular cavities and the impact factors for cavities formation is much complicated, but some evidence of creep damage can be seen in substrucure. The shapes of the laths are changed; in particular, the lath boundaries look like bamboo knots, called cell structure, which is a typical microstructure morphology caused by creep. Many low dislocation density regions appeared in the lath structure, and some typical substructures can be seen in Fig.9. The substructure seems to develop as subgrains boundaries are formed by dislocation movement during the creep process. A significant reduction of dislocation density is observed, and few dislocation-free regions can be seen. Extensive carbide precipitates can be seen at prior austenite and martensite lath boundaries, with the finer precipitates in martensitic laths. Large coarsening carbides in irregular spheroid formed along the boundaries. Compared with virgin material, the carbide morphology coarsened distinctly. The observations indicate that the matrix of the tempered martensite has undergone a deterioration during long-term creep. The dislocations climbed or glided and terminated at boundaries. As the number of dislocations at the boundaries increased, networks formed and substructures developed. The carbides morphology in boundaries coarsened distinctly, and most of the strengthening phase have dissolved or coarsened.

Except the coarsening characteristic, martensitic steels have shown common evolution in microstructure is martensite decomposition and substructure change during long-term exposure at high temperature. It is commonly accepted that martensitic structure under a degeneration with creep exposure prolonging. The dislocation density decreases and With the precipitates coarsening and the density of fine particles within the matrix decreases during long-term exposure, the martensitic lath boundaries is indistinct or even disappeared which undergone a degeneration, and coarsening of laths is processing by two ways [68,69,70], the recovery of dislocations within lath boundaries and the recombination of two subgrain boundaries which mainly takes place near the triple point of lath boundaries by moving and causes the disappearance of lath boundaries. Forth more, during moving of lath boundaries to cause progressive local-coalescence, dissolution and re-precipitation of $M_{23}C_6$ carbides distributing along lath boundaries take place repeatedly. Observations did not confirm that the MX precipitates evolve during creep or thermal exposure. The solid solution strengthening effect by alloy elements maybe weaken but was found to be negligible. The precipitation of Laves phases should not strongly affect resistance of this steel to creep deformation, however, that damage cavities were often found next to Laves phases, so that Laves phases could affect the resistance of that steel to creep fracture.

Generally it is accepted that the increase of Cr content in $M_{23}C_6$ precipitates for all ferritic steels, and the increase of Mo content in precipitate in eutectoidic carbide as M_3C or as M_6C. The composition variation in precipitates and interparticle distance decrease with carbides coarsening are believed as eight indexes of creep exposure and microstructural degradation. However, there is a notable difficulty in correlation of microstructural evolution (such as second phases) to residual life assessment. It is not only necessity to know the actual state of virgin material that for the same type of material can significantly vary from heat to heat or product to product, but the exposure condition and history are also the basis factors to affect the evolution course. Furthermore, there is no quantitative relationship has been established to correlate the parameters of microstructural evolution with life depletion.

The primary difficult for evaluation the life of components with the causation for the dagradation of heat-resistant steels

4.2 Microvoids formation in grain boundaries

The concept of microvoid or micro-cavity formation at grain boundaries has been studied and developed in the1970's and commonly recognized and applied in all European countries with the Neubauer classification and derived methods for decades [71]. The principle is based on the fact that creep evolution of heat resistant steels is related to the appearance of cavities some time before rupture. These cavities gradually form microcracks by interlinkage and at the end come to initiate the rupture. So the emergence of microvoids means that the materials in service damaged seriously.

Assessment class	Structural and damage conditions
0	As received, without thermal service load
1	Creep exposed, without cavities
2a	Advanced creep exposure, isolated cavities
2b	More advanced creep exposure, numerous cavities without preferred orientation
3a	Creep damage, numerous orientated cavities
3b	Advanced creep damage, chains of cavities and/or grain boundary separations
4	Advanced creep damage, microcracks
5	Large creep damage, macrocracks

Table 10. Neubauer schematic assessment of the microstructure

Table 10 is a revision of Neubauer classification is presented in the German VGB "Guidelines for the assessment of microstructure and damage development of creep exposed materials for pipes and boiler components", which is considered as one of the most updated reference document in Europe. Another important aspect of correlation of cavities presence to creep progress can be found in the Neubauer documents, where it is stated that "a noticeable cavity formation takes place at grain boundary at the end of secondary creep". A graphical representation is shown in Fig.10. Although above literature examples of damage grade allocation seem to not completely congruent with the creep curve, it can be considered that grade 4 and 5 should be taken as representative of different stages of tertiary creep, while grade 3 figured as the transition point among secondary and tertiary and grade 2 considered as representative of secondary creep.

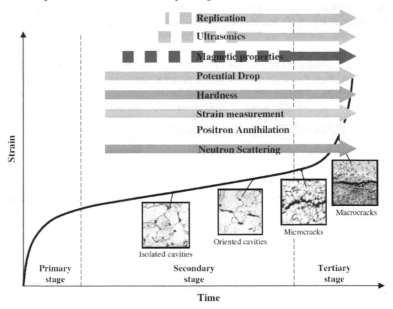

Fig. 10. Typical creep curve with evolution of microstructural damage, and various NDT techniques with indication of applicability ranges [from ref.72].

An interesting summarising is the correlation of microcavities damage and residual life assessment, which is one of the methods followed by some researchers and based on the analysis of experimental results. Graph of the summarising correlation of damage level and expended life fraction is presented in Fig.11. Other reviewers [73,74,75] gave much more critical summary of cavitation. A cavitation derived parameter named A-parameter (number fraction of cavitated grain boundaries) have been used in some studies related to low alloy ferritic steel for pipe and rotor.

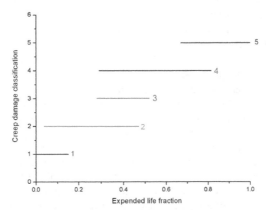

Fig. 11. Creep Damage Classification and Expended Life Fraction

Generally the size and density of the cavities increase as creep progresses, which are also dependant on material type, however it is accepted commonly that the microvoids formation and evolution figure is a prominent index to estimate damage degree.

As to the theoretical research about microvoid formation, past studies[75,76] indicate that cavity nucleation is associated with grain boundary and second phase particles in it, and the presence of surface active elements such as P and Sn in grain boundary makes cavity nucleation easier. The density of cavities (number of cavities per unit grain boundary area) increases with creep exposed time and temperature, and the applied stress enhances the cavity density too. Cavities growth is controlled by two mechanisms: diffusion growth and constraint growth, the former is dominating at high stress level, and the latter one is dominating at much lower stresses.

The size of microvoid is in the range of micron and it cannot be detected by conventional NDT techniques, so metallographic investigation is required to observe it morphology and distribution.

5. Techniques for the residual life assessment of components in power plant

Assessing the integrity and stability of high-temperature, service-exposed components is a interesting topic of for power plant users. The primary reason for microstructure damage and mechanical property deterioration is the degressive creep behavior in long-term exposure at elevated temperature. The residual life assessment is a important procedure in routine examination.

5.1 Inspection method on in-service components

Except the various NDT techniques given in fig.10, the technology of investigation microstructure to correlate the material microstructural evolution and creep exposure have been performed is metallographic observation of the creep specimen after tests, and the microstructural investigation on actual in service components is usually gained with replica technique.As a non destructive technique, replica can be assumed e or at least very low invasive technique, and this is the first requirement for an investigation applied for residual life evaluation.

The replica technique is essentially application of the metallographic specimen preparation (grinding, polishing, and etching) to a limited area to investigate on the examined component, then the reproduction of the prepared surface on a thin foil of polymeric material. The reproduction of prepared surface on plastic material is achieved by softening of polymeric thin foil with adequate solvent and then hardening the plastic foil due to solvent evaporation. Removed replica can be observed by optical microscope. With further preparation by coated a conductive support (carbon or gold), replica can be observed with scanning electron microscope achieving the possibility to observe reproduced microstructure aspect with higher magnification. If microstructure evolution is the goal, removing a layer of surface (about 0.3-0.5mm thickness reduction) is recommended to avoid the external layer of oxidation or decarburized material.

In order to observation and identification of the second phases in the examined material, an extractive replica should be prepared. It is prepared following the same procedure of above morphological replica preparation, in addition to these the critical for the extractive replica is etching the surface to remove the metal matrix without any perturbation on carbides and other precipitates. The extractive replica removed from the examined surface contains the precipitates that can be observed late with electron microscope (SEM). Alternative techniques for carbon extraction replica are available analyzed by TEM. The sample is prepared by transferring the carbides and precipitates from polymeric resins (dissolved) to a thin layer of graphitic carbon.

Today, metallographic replica testing is the most common inspection method for creep damage monitoring in piping systems and steam boilers. As ductile materials, 9-12Cr martensites show that cavity formation is less distinct, and this process starts very late. Furthermore, cavity formation investigations show some differences between the various heat resistant steels, even difference between various batches of the same material. So their creep damage propagation needs to be characterized by additional methods. Classical replica testing are the small inspection area and the analysis limited to the outer surface only, without information on damage within the wall. For this purpose other methods of condition monitoring need to be found.

US-laminography represents an advanced test method for damage detection, which allows to inspect larger areas [77]. Creep deformation or creep strain measurements by capacitive gauges represent another traditional creep expansion measurement using special "warts" or "pips" at the outer surface to measure diameter or circumference, which is a costful effective method. Thus, extensive efforts have to be undertaken to investigate these materials and material degradation. The classical replica test will remain one method of creep damage monitoring in the future. However, it needs to be accompanied by other qualified methods for ductile materials like 9Cr steels. As mentioned above, past studies show that martensitic structure and precipitates coarsen obviously in 9-12Cr ferritic steels during long-term exposure at high temperature. Meanwhile, the mechanical properties present variation, ex-service materials incline to embrittle at room temperature, and the combination of phosphorus segregation and carbide coarsening at grain boundary does account for the embrittlement tendency [78]. However, more representative characteristics or parameters have to be proposed. The residual life maybe predicated from consumed life by sophisticated calculations, using material data from NDT and/or available data from the as received steel condition. Such as Z-Factor-Method is one of the potential procedures [79].

5.2 Life assessment methodology for components

It is generally agreed that at present the only valid technique for quantitative evaluation of damage to service-exposed material is the creep-rupture testing of the service exposed samples. Hence, a more accurate and reliable residual life assessment (RLA) of power plant components is strictly related to the improvement of this technique. Moreover, various reasons, for instance, the high cost of restoration of the sampled zone in the inservice component, the induced long times of plant shut-down and the occasional inability to sample specific components or their localized areas, have influenced the tendency towards non destructive methods.

For the integrity analyses of plants and residual life studies, the analysis technology is summarized in some papers [87,80,81], which usually combines the following integrated life assessment methodologies:

1. Finite element analysis to define component loading situation (stress and temperature fields);
2. Mechanical tests to define the material's response in the real service aging condition;
3. NDT to define existence of cracks, metallographic investigation to estimate microstructure evolution;
4. Computer codes to assess component integrity and to calculate residual lives.

Advanced codes of integrity and life assessment as well as some cases are proposed by electric companies or research institutes, such as ENEL, Electric Power Research Institute (EPRI) and

CISE etc. The code damage analysis is based on appropriate materials data. They also provide advanced test methodology assessment and verifications of damage analysis methods.

The present methodology for estimating residual creep life involves accelerated iso-stress creep-rupture tests at operative temperature or above on service-exposed material, and then the life predictions are estimated by extrapolations of test results.

Therefore, the operative stress field in the component must be calculated firstly in order to determine the stress utilized during the iso-stress creep-rupture tests.

Based on the metallographic examination and finite element analysis, experimental data combine with suitable calculations will enable an integrated access for estimating the remaining life of the imitation components with more accurate and realistic forecast of failure.

5.3 Summary and comments

VGB guidelines for the assessment of microstructure and damage development in power plant components is considered commonly in worldwide, and this guidelines are accepted as the most updated reference document for microstructure and residual life assessment correlation.

For residual life assessment of power plant components, different materials have their own degradative characteristics, so it is important to remark that all other, except cavities, microstructural observed parameters need to be evaluated on the basis of the as received material correspondent status in order to avoid any misleading deduction from generic or recommended microstructure variation not actually correlated to service exposure. Particularly the 9-12Cr martensites are complex metallurgically, they need careful control of heat treatment to avoid δ-ferrite formation and ensure transformation to martensite fully. During tempering stage, the conventional balance between strength and ductility is required. When components manufactured, the processes, such as bending and welding, must follow the strict control procedures. Some investigations reveal that the some defects, like microvoids, may formation during these processes [82]. Many reports about components failure in power plants show that the primary causations are related to inadequate heat-treatment or welding defects. Some theoretical researches [83] reveal that these improper processes induce abnormal microstructure and lead to creep strength deterioration.

It is believed that the life of equipments in power plants is closely related with the microstructure revolution. In other words, microstructural evolution in exposed to service materials is a key tool for a correct evaluation of material status and allowable service extension. However, with the complicated reasons mentioned above, reliable life assessment shouldn't be made only by means of microstructure inspections, it is preferable that together with other inspections and mechanical tests are included. For the widely applied ferritic low alloyed steels, the evaluation of micro-cavitation presence and creep damage evolution seems to be the most consolidated approach. For every other microstructural aspect (except microcavities) to be monitored, it is very important that the evaluation is made by comparison with the original status in virgin material. Some materials microstructure and their evolution are very sensitive to their factual statuses and exposure conditions.

From the view point of the techniques applied for monitoring microstructure on in-service components, replica are surely a consolidated and reliable technique, attention should be paid in any case also to alternative low invasive techniques for sampling as the ones applied for small punch or impression creep specimen preparation.

6. Acknowledgments

The author would like to thank the support from the National Science Foundation of China under contract no.50760176 for the correlative research project.

7. References

[1] I. P. Seliger, I. U. Gampe. Life Assessment of 9Cr Steel Components OMMI (Vol. 1, Issue 2) August 2002.

[2] Residual life assessment and microstructure, ECCC Recommendations –Volume 6 [Issue 1], 2005.

[3] A. D. Gianfrancesco, D. Venditti, D. J. Allen, et al. Applications of advanced low-alloy steels for new high-temperature components, Office for Official Publications of the European Communities, ISBN 978-92-79-10006-2. Luxembourg 2009.

[4] M. Wada, K. Hosoi and O. Nishikawa. Atom probe analysis of 2.25 Cr1Mo steel. Acta Metallurgica Volume 30, Issue 5, May 1982, pp.1013-1018 .

[5] J. Purmenský, V. Foldyna, Z. Kuboň, Proceedings of the 8-th International Conf. on Creep and Fracture of Engineering Materials and Structures, Tsukuba, Japan,1999, p. 419.

[6] Miyata K, Sawaragi Y. Effect of Mo and W on the phase stability of precipitates in low Cr heat resistant steels. ISIJ Int., 2001, 41, 281.

[7] Yamada K, Igarashi M, Muneki S, Abe F. Effect of Co addition on microstructure in high Cr ferritic steels. ISIJ Int., 2003, 43, 1438.

[8] J. Janovec, A. Výrostková and M. Svoboda. Metall. Trans., 1994, 25A, 267.

[9] B.A. Senior. Matr. Sci. Eng. A, 1988, 103, 263.

[10] A. Výrostková , A. Kroupa, J. Janovec and M. Svoboda. Acta Mater., 1998, 46(1), 31.

[11] A Strang, V.Vodarek. Z phase formation in martensitic 12CrMoVNb steel. Mater Sci Technol., 1996,12,552.

[12] H.J.Goldschmidt, et al., J. Iron Steel Inst., 1948,160,345

[13] D.A Porter, K.E. Easterling. Phase transformation in metals and alloys. 2nd ed. Chapman & Hall; 1992. p. 420.

[14] Hu Zheng-Fei Zheng-Guo Yang, et al. An investigation of the embrittlement in X20CrMoV12.1 power plant steel after long-term service exposure at elevated temperature, Material Sci & Eng. A, 2004,383,224-8.

[15] M Hättestrand and H-O Andrén. Evaluation of particle size distributions of precipitates in a 9% Cr-steel using energy filtered transmission electron microscopy. Micron, 2001,32,789.

[16] Hu Zhengfei, Yang Zhenguo. Identification of the precipitates by TEM and EDS in X20CrMoV12.1 for Long-term Service at elevated temperature, J.Mater. Eng. & Perform., 2003 , 12(1), 106-11.

[17] A Golpayegani and H-O Andrén. Mechanism for beneficial effect of boron addition on creep resistance of 9-12% chromium steels, In "Solid-to-Solid Phase Transformations in Inorganic Materials 2005", Eds. JM Howe, LE Laughlin, JK Lee, U Dahmen and WK Soffa. TMS, Warrendale PA, 2005, Vol. 1 pp. 461.

[18] F. Abe, H. Semba, T. Sakuraya. Effect of Boron on Microstructure and Creep Deformation Behavior of Tempered Martensitic 9Cr Steel. Materials Science Forum, 2007, 539-543, 2982-7.

[19] V. Thomas Paul, S. Saroja, M. Vijayalakshmi. Microstructural stability of modified 9Cr–1Mo steel during long term exposures at elevated temperatures. Journal of Nuclear Materials 2008,378, 273–281.

[20] Takashi Onizawa, Takashi Wakai, Masanori Ando and Kazumi Aoto. Effect of V and Nb on precipitation behavior and mechanical properties of high Cr steel Nuclear Engineering and Design, 2008, 238(2), 408-416.

[21] J. Hald. Microstructure and long-term creep properties of 9–12% Cr steels. International Journal of Pressure Vessels and Piping, 2008, ,85, 30.

[23] Y. Hosoi, N. Wade, S. Kunimitsu, T. Urita. J Nucl Mater 1986, 461,141.

[24] L. Korcakovaa, J. Hald, M. A.J. Somers. Materials Characterization, 2001,47, 111.

[25] R.U. Husemann , W. Bendick, K. Haarmann. The new 7CrMoVtiB10-10 (T24) material for boiler water walls. PWR-Vol. 34, 1999 joint power generation conference, vol. 2. New York: ASME; 1999. p. 633–40.

[26] J.Arndt, K. Haarmann, G. Kottmann, J.C. Vaillant, W. Bendick, F.Deshayes. The T23/T24 book—new grades for waterwalls and superheaters. Vallourec & Mannesmann Tubes; 1998

[27] W. Bendick, J. Gabrel, B. Hahn, B. Vandenberghe. New low alloy heat resistant ferritic steels T/P23 and T/P24 for power plant application International Journal of Pressure Vessels and Piping, 2007, 84, 13.

[28] V.K. Sikka, R.L. Klueh, P.J. Maziasz,, et al. Mechanical properties of new grades of Fe–3Cr–W alloys. Amer Soc Mech Eng Pressure Vessel Piping, 2004, 476: 97.

[29] M. Igarashi. 2.25Cr–1.6W–V–Nb steel. Creep properties of heat resistant steels and superalloys. In: Yagi K, Merkling G, Kern TU, Irie H, Warlimont W, editors. Landolt–Bornstein Numerical Data and Functional Relationships in Science and Technology, Group VIII: Advanced Materials and Technologies, vol 2. Berlin, Heidelberg, New York: Springer-Verlag; 2004. pp. 74–83.

[30] J. Yu, C.J. McMahon, Metall. Trans., 1980,11A, 277.

[31] S.-H. Song, H. Zhuang , J. Wu, L.-Q. Weng, Z.-X. Yuan, T.-H. Xi. Dependence of ductile-to-brittle transition temperature on phosphorus grain boundary segregation for a 2.25Cr1Mo steel. Materials Science and Engineering A, 2008, 486, 433..

[32] J.D. Robson, H.K.D.H. Bhadeshia. Modelling Precipitation Sequences in Power Plant Steels. Part 1: Kinetic Theory. Materials Science and Technology 1997,13: pp. 631 – 639.

[33] Singh, R., Singh, S. R. Remaining Creep Life Study of Cr-Mo-V Main Steam Pipe Lines, Int. J. Pres. Ves. &Piping 73 1997, 89–95.

[34] N. Fujita, H.K.D.H. Bhadeshia. Modelling Simultaneous Alloy Carbide Sequence in Power Plant Steels ISIJ International, 2002, 42 (7), 760.

[35] A. Baltušnikas, R. Levinskas, I. Lukoštūtė. Kinetics of Carbide Formation During Ageing of Pearlitic 12X1MΦ Steel. Mater. Sci., 2007,13(4),286-292.

[36] J.D. Robson, H.K.D.H. Bhadeshia. Kinetics of Precipitation in Power Plant Steels Calphad, 1996, 20 (4) ,447.

[37] T. Fujita. Advances in 9–12%Cr heat resistant steels for power plant. In: Viswanathan R, Bakker WT, Parker JD, editors. Proceedings of the 3rd Conference on Advances in Material Technology for Fossil Power Plants. London (UK): The Institute of Materials; 2001. p. 33–65.

[38] K.H.Mayor, W.Bendick, R.U.Husemann, T.Kern and R.B.Scarlin, New Materials for Improving the Efficiency of Fossil-field Thermal Power Stations, 1988 Inter. Joint Power Generation Conf., PWR-Vol.33, Vol.2, ASME 1988, in Baltimore, USA

[39] Super 12 Cr steels, an update, Climax Molybdenum Company, M571 782 15M, New York, 1983.

[40] V. Vodárek and A. Strang. Effect of nickel on the precipitation processes In 12CrMoV steels during creep at 550℃. Scripta Materialia, 1998, 38(1), 101–106.

[41] P. J. Grobner And W. C. Hagel, The Effect of Molybdenum on High-Temperature Properties of 9 Pct Cr Steels. Metal Trans., 1980,11A, 633-642.

[42] M. Ohgami, H. Mimura, H. Naoi, T. Fujita, in: Proceedings of the Fifth International Conference on Creep of Materials, 1992, p 69.

[43] A. Iseda, Y. Sawaragi, S. Kato, F. Masuyama, in: Proceedings of the Fifth International Conference on Creep of Materials, 1992, p. 389.

[44] K. Niederhoff, Werkstoffechniche Besonderheiten beim Schwe en Warmfester Martensitischer Cr-Staele fur den Kraftwerksbau, Duisburg, VGB.

[46] C. Berger, A. Scholz, Y.Wang, K.H. Mayer, Z. Metallkd. 2005, 96, 668–674.

[47] K.H. Mayer, Proc. of 29. MPA Seminar Materials & Component Behaviour in Energy & Plant Technology Stuttgart 9–10, October 2003, MPA Stuttgart, Germany, 2003.

[48] R.W.Vanstone. Microstructure in Advanced 9-12%Cr Steam Turbine Steels, in Quantitative Microscopy of High Temperature Materials, eds A Strang and J Cawley, IOM Communications, London 2001 pp 355-372.

[49] T. Horiuchi, M. Igarashi, and F. Abe. Iron Steel Ist. Jpn. Int., 2002,vol. 42, Suppl., p. S67.

[50] P. Lacombe, B. Baroux, G. Beranger (ed.), Stainless Steels, Les Ulis, Editions de Physique (1993)

[51] R. Blum, R.W. Vanstone, C. Messelier-Gouze, Proceedings 4th International Conference on Advances in Materials Technology for Fossil Power Plant, 2004,p116.

[52] B. Rudolph, R.W. Vanstone. Proceedings 4th International Conference on Advances in Materials Technology for Fossil Power Plant, 2004, p317.

[53] R. Rautio, S. Brua. Proc. 4th Int. Conf. on Advances in Materials Technology for Fossil Power Plant, 2004, 274.

[54] High Temperature Grades, Data Sheet. Sandvik Materials Technology, 2000.

[55] B. Nath and F. Masuyama, "Materials Comparisons between NF616, HCM12A and TB12M – 1: Dissimilar Metal Welds", New Steels for Advanced Plant up to 620 °C, ed E. Metcalfe, PicA, Drayton, Oxfordshire, 1995, pp 114-134.

[56] Jian Cao, Yi Gong, Zhen-Guo Yang, Xiao-Ming Luo, Fu-Ming Gu, Zheng-Fei Hu. Creep fracture behavior of dissimilar weld joints between T92 martensitic and HR3C austenitic steels. International journal of pressure vessels and piping, 2011, 88(2-3), 94.

[57] P. Mohyla, V. Foldyna. Improvement of reliability and creep resistance in advanced low-alloy steels. Materials Science and Engineering, 2009,A 510–511, 234.

[58] DL/T 773-2001, Spheroidization evaluation standard of 12Cr1MoV steel used in power plant, 2002.

[59] M. Drew, S. Humphries, K. Thorogood, N. Barnett.Remaining life assessment of carbon steel boiler headers by repeated creep testing. Int. J. Press. Ves. & Pip., 2006, 83,343.

[60] Linee guida per la valutazione di vita residua per componenti in regime di scorrimento viscoso ISPESL (Istituto Superiore per la Prevenzione e la Sicurezza del Lavoro) Circ. 48/03

[61] K.R.Williams. Microstructural examination of several commercial ½Cr½Mo¼V casts in the as received and service exposed conditions, CEGB research laboratory report 1979

[62] A. Benvenuti, N. Ricci, and G. Fedeli. Microstructural changes in long term aged steels Microstructures and mechanical properties of ageing material –The mineral, metals, & materials society 1993.

[63] A. Benvenuti, N. Ricci, and G. Fedeli. Evaluation of microstructural parameters for the characterisation of 2½ Cr 1Mo steel operating at elevated temperature. Proceeding of Swansea conference 1990.

[64] P.Battaini, G.Marino, J.Hald. Interparticle distance evolution on steam pipes 12%Cr steel during power plants' service time Proceeding of Swansea conference 1990.

[65] J.P. Ennis. Creep strengthening mechanisms in high chromium steels. In: Bakker, W.T., Parker, J.D. (editors). Proceedings of the Third Conference on Advances in Materials. Technology for Fossil Power Plants. London (UK): The Institute of Materials, 2001,pp187–194.

[66] Sawada, K., Maruyama, K., Hasegawa, Y., Muraki, T.: Creep life assessment of high chromium ferritic steels by recovery of martensitic lath structure. Proceedings of Eighth International Conf. on Creep and Fracture of Engineering Materials and Structures, Tsukuba, Japan, 1999,Nov. 1–5.

[68] F. Abe. Coarsening Behavior of Lath and Its Effect on Creep Rates in Tempered Martensitic 9Cr-W Steels [J]. Mater. Sci. & Eng., 2004, 387A: 565.

[69] C.G. Panait, A. Z.-Lipieca, T. Koziel et al. Evolution of dislocation density, size of subgrains and MX-type precipitates in a P91 steel during creep and during thermal ageing at 600°C for more than 100,000h. Materials Science and Engineering. 2010.A527,4062.

[70] C. Panait, W. Bendick, A. Fuchsmann, A.-F. Gourgues-Lorenzon, J. Besson, in: I.A. Shibli, S.R. Holdsworth (Eds.), Proc. Creep & fracture in high temperature components: Design & life assessment issues, DEStech Publications ISBN978-1-60595-005-1, Lancaster, U.S.A, 2009, pp 877–888.

[71] Criteria for prolonging the safe operation of structures through the assessment of the onset of creep damage using non-destructive metallographic measurements (Neubauer/TÜV) reprint from "Creep and fracture of engineering materials and structures" march 1981.

[72] G. Sposito, C.Ward, P.Cawley, P.B.Nagy, C.Scruby, A review of non-destructive techniques for the detectionof creep damage in power plant steels. NDT&E International,2010, 43,555.

[73] N.G. Needham, Cavitation and fracture in creep resisting steels. EUR 8121 EN, Commission of the European Communities, Directorate-General Information Market and Innovation, Luxembourg,1983

[74] G.Lyell, A.Strang, D.J.Gooch, et al. Techniques for remanent life assessment of 1CrMoV rotor forging Cavitation assessment. National Power Confidential note 1990.

[75] K.Saito, A.Sakuma, M.Fukuda, Recent Life Assessment Technology for Existing Steam Turbines, JSME International Journal Series B,2006, Vol.49, pp.192-197.

[76] P. Shewmon, P. Anderson, Void nucleation and cracking at grain boundaries, Acta Materialia, 1998, V 46(14), 4861.

[77] K.G. Schmitt-Thomas, E.Kellerer, E.Tolksdorf. Zerstoerungsfreies Pruefverfahren zum Nachweis von Zeitstandschaedigung mit Hilfe von Ultraschall-Oberflaechenwellen. VGB Konferenz Werkstoffe und Schweisstechnik im Kraftwerk, Cottbus, 1996.

[78] Hu Zheng-Fei et al, An investigation of the embrittlement in X20CrMoV12.1 power plant steel after long-term service exposure at elevated temperature, Material Sci & Eng. A, 2004,383,224.

[79] B.Melzer, P.Seliger, W. Illmann. Verbesserte Lebensdauerabschaetzung kriechbeanspruchter Rohrbogen mittels bauteilspezifischer Kennwerte. VGB Kraftwerkstechnik,1993, 73, 4.

[80] V. Bicego, E.Lucon, R.Crudeli, Integrated technologies for life assessment of primary power plant components. Nuclear Engineer and Design, 1998, 182,113.

[81] A. Garzillo, C.Guardamagna, L.Moscotti, L.Ranzani. Int. J. Pres. Ves. & Piping, 1996, 66, 223.

[82] B.A.Senior, F.W.Noble, Annealing behavior of deformation-induced voids in 9Cr1Mo steel. Mater. Sci. & Tech., 1985, 1, 968.

[83] D.R.Barraclough, D.J.Gooch, Effect of inadequate heat treatment on creep strength of 12CrMoV steel. Mater. Sci. & Tech., 1985, 1, 961.

12

Spectrophotometric Determination of 2-Mercaptobenzothiazole in Cooling Water System

Fazael Mosaferi, Farid Delijani
and Fateme Ekhtiary Koshky
*East Azarbayjan Power Generation
Management Company, Tabriz,
Iran*

1. Introduction

Metals can get into cooling system water from corrosion of the materials used to construct the equipment (cooling tower, heat exchanger, and piping), or from the use of conditioning chemicals containing metals. Copper is a common material of construction in cooling systems. Because of its excellent heat transfer efficiency, heat exchangers, and condensers are often made from copper. Copper piping is also commonly found in cooling systems. Moreover, Copper is metal that has a wide range of applications due to its good properties. It is used in electronics, for production of wires, sheets, tubes, and also to form alloys. Copper is resistant toward the influence of atmosphere and many chemicals, however, it is known that in aggressive media it is susceptible to corrosion. The use of copper corrosion inhibitors in such conditions is necessary since no protective passive layer can be expected. The possibility of the copper corrosion prevention has attracted many researchers so until now numerous possible inhibitors have been investigated. Amongst them there are inorganic inhibitors [1], but in much greater numbers there are organic compounds and their derivatives such as azoles, amines, amino acids [2,3] and many others. It is noticed that presence of heteroatoms such as nitrogen, sulphur, phosphorous in the organic compound molecule improves its action as copper corrosion inhibitor. This is explained by the presence of vacant d orbitals in copper atom that form coordinative bonds with atoms able to donate electrons. Interaction with rings containing conjugated bonds, electrons, is also present. Based on these results more and more compounds containing numerous heteroatoms and functional groups are developed synthesized since it is noticed they are responsible for good properties regarding corrosion inhibition because they enable chemisorption.

2. Inorganic copper corrosion inhibitors

The use of inorganic inhibitors as an alternative to organic compounds is based on the possibility of degradation of organic compounds with time and temperature. Three different inorganic inhibitors are investigated: chromate CrO_4^{2-}, molybdate MoO_4^{2-} and tetraborate

$B4O_7^{2-}$ in concentration of 0,033M in solution containing 850g/l LiBr and has pH 6,9. Chromate is generally accapted as efficient corrosion inhibitor that can passivate metals by forming a monoatomic or polyatomic oxide film at the electrode surface, but it is also known that it can promote corrosion acting as a cathodic reactive. [1]

3. Organic copper corrosion inhibitors

3.1 Amines

Copper corrosion inhibition in de-aerated, aerated, and oxygenated HCl [4] and NaCl [5] solutions by N-phenyl-1,4-phenylenediamine (NPPD) is investigated. The NPPD adsorbs on the copper surface whereat Cu undergoes oxidation to Cu^+ and form insoluble complex Cu^+-NPPD on the surface. The efficiency increases with time and inhibitor concentration.

The behavior of secondary amines as copper corrosion inhibitors in acid media, 0.5M hydrochloric acid and 0.5M sulphuric acid, is studied. [6] The homologous series of aromatic secondary amines with various substituents is investigated.

3.2 Amino acids

Amino acids form a class of non-toxic organic compounds that are completely soluble in aqueous media and produced with high purity at low cost. These properties would justify their use as corrosion inhibitors.

J.B.Matos [2] studied the effect of cysteine (Cys) on the anodic dissolution of copper in sulfuric acid, at room temperature using electrochemical methods. Cys ($HSCH_2CHNH_2COOH$) contains three dissociable protons, and in aqueous solutions ionization depends upon pH. Acording to the copper dissolution mechanism proposed for sulfate media in the absence of cys the main species present on the copper surface at low overpotentials is the intermediate Cu(I)ads.

3.3 Triphenylmethane derivatives

Two nitrogen containing organic compounds which are triphenylmethane (($C_6H_5)_3CH$) derivatives, fuchsin basic FB (rosaniline chloride) (C20H19N3•HCl) and fuchsin acid sodium salt FA($C_{20}H_{17}N_3O_9S_3Na_2$), are tested as new copper corrosion inhibitors. [7,8] These compounds are thought to be good candidates due to the presence of chloride ion in FB and the polar or charged nature of the more complex FA surfactant molecule.

3.4 Thiole group compounds

The inhibition of copper corrosion in 1.5% NaCl solution is studied at 25, 35 and 45°C using three inhibitors: thiosemicarbazide (inh 1), phenyl isothiocyanate (inh 2) and their condensation product 1-phenyl-2,5-dithiohydrazodicarbonamide (inh 3). [9] It is concluded that all the three compounds are efficient corrosion inhibitors whereat the inhibition efficiencies follow the sequence: inh3>inh1>inh2. The enhanced effectiveness of the inh 3 can be correlated with the structure and size of molecule, inh 3 has four nitrogen atoms, two sulphur atoms and delocalised ð electron density acting as active centres and the largest surface area. Mechanism of inhibition is proposed as adsorption only in case of inh 2; inh 1 at lower concentrations inhibits corrosion through adsorption while at higher concentrations Cu(I) complex is formed that gradualy oxidises into Cu(II) complex.

3.5 Phosphates as copper corrosion inhibitors

Copper corrosion by-product release to potable water is a complex function of pipe age, water quality, stagnation time and phosphate inhibitor type. Phosphates can be phosphoric acid, combination of orthophosphoric acid and Zink orthophosphate, polyphosphate or blend of orthophosphoric acid and polyphosphate. It is noticed [10] that dosing of 1mg/1 orthophosphate led to reductions in copper release ranging from 43-90% when compared to the same condition without inhibitor regardless of pipe age, water quality or stagnation period. Ortophosphate and hexametaphosphate have beneficial effects on copper release, but ortophosphate leads to greater reductions in copper release when compared to hexametaphosphate.

3.6 Azoles

Azoles are organic compounds containing nitrogen atoms with free electron pairs that are potential sites for bonding with copper and that enable inhibiting action. Also, there is a possibility of introduction of other heteroatoms and groups in molecules of these compounds so there is a wide range of derivatives that exhibit good inhibition characteristics.

El-Sayed M.Sherif [11-14] investigated the influence of 2-amino-5-ethylthio-1,3,4-thiadiazole (AETD) on copper corrosion in aerated HCl solution [11] as well as the influence of 2-amino-5-ethylthio-1,3,4-thiadiazole (AETD) [12], 2-amino-5-ethyl-1,3,4-thiadiazole (AETDA) [13] and 5- (phenyl)-4H-1,2,4-triazole-3-thiole (PTAT) [14] in NaCl solution. It is expected that these compounds show high inhibition efficiency since they are heterocyclic compounds containing more donor atoms, besides that they are non-toxic and cheap.

3.6.1 Benzotriazoles

In recent years, investigators have shown that a system of tarnish or corrosion control for copper, brass and bronze can be built around the organic compound, 1, 2, 3, benzotriazole. Benzotriazole forms a strongly bonded chemisorbed two-dimensional barrier film less than 50 angstroms thick. This insoluble film, which may be a monomolecular layer, protects copper and its alloys in aqueous media, various atmospheres, lubricants, and hydraulic fluids. Benzotriazole also forms insoluble precipitates with copper ions in solution (that is, it chelates these ion), thereby preventing the corrosion of aluminum and steel in other parts of a water system.

3.6.1.1 Inhibition mechanism of benzotriazole

Benzotriazole (BTA), whose structure is shown in Figure 1, has been used for a long time as an important corrosion inhibitor for Cu and its.

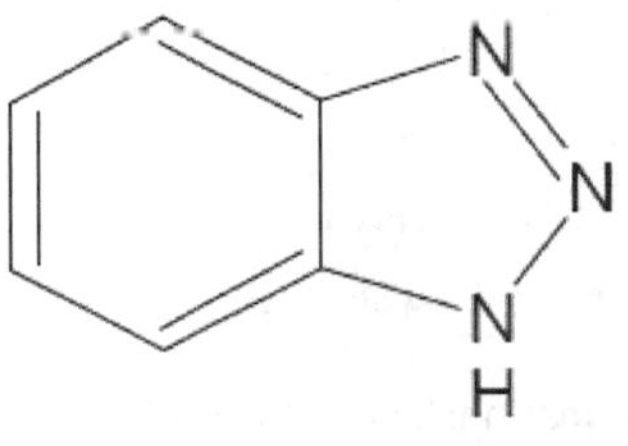

Fig. 1. Chemical structure of benzotriazol (BTA)

The protective barrier layer, which consists of a complex between Cu and BTA molecules can be formed by the immersion of the Cu surface in a solution of BTA or by vapor transport from impregnated paper or electrochemically . This barrier is insoluble in water and many organic solvents and grows with time to a certain thickness depending on the BTA concentration and the pH of the solution. [15]

3.6.2 Benzothiazole (BT)

Benzothiazole enter the environment from a variety of sources such as the leaching of rubber products, but particularly by routes associated with the manufacture and use of mercaptobenzothiazole (MBT) and MBT-based rubber additives, fine particles of automobile tires, and antifreeze. All benzothiazole used are solids at room temperature with the exception of benzothiazole (BT), which is liquid.

Benzothiazoles form a part of xenobiotic, heterocyclic, molecular structures comprising a benzene ring fused with a thiazole ring. Their general structure is shown in Figure 2.

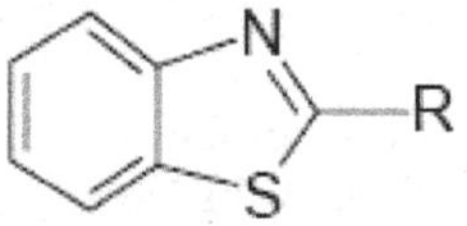

Fig. 2. General structure of benzothiazoles

Table 1 shows that BT possesses a high solubility (4300 mg/L), this is probably due to its high polarity and the fact that it is a liquid at room temperature. BT is also considered as volatile with a vapour pressure 0.0143 mm Hg (25°C).

R =	formula	Chemical names	Abreviation	Molecular weight	Water Solubility (mg/L) 25°C	Octanl/water (log₁₀ K$_{ow}$)
SO3H	$C_7H_5N\ O_3S_2$	Benzothiazol Sulfonic acid	BTSA	215	nf	-0.99
OH	C_7H_5NOS	Hydroxybenzothiazole	OHBT	151	2354 [c]	2.35 [b]
NH2	$C_7H_6\ N_2S$	Aminobenzothiazole	ABT	150	310.3 [c]	2.00 [d]
SH	$C_7H_5\ N\ S_2$	Mercaptobenzothiazole	MBT	167	120 *,[a]	2.86 [f]
H	C_7H_5NS	Benzothiazole	BT	135	4300 [a]	2.17 [b]
CH3	C_8H_7NS	Methylbenzothiazole	MeBT	149	366.3 [c]	2.72 [d]
SCH3	$C_8H_7\ N\ S_2$	Methylthiobenzothiazole	MTBT	181	125 *,[5]	3.22 [g]
SCH2SCN	$C_9H_6\ N_2S_3$	Thiocyanomethylthiobenzothiazole	TCMTB	238	125 *,[a]	3.12 [a]
S-S-BT	$C_{14}H_8\ N_2S_4$	Dithiobisbenzothiazole	MBTS	332	10 [a]	4.66 [d]

24°C, nf = not found. [a] CHEM INSPECT TEST INST (1992). [b] Hansch, C et al. (1995). [c] Meylan, WM et al. (1996). [d] Meylan, WM &Howard, PH (1995). [e] Brownlee, BG et al. (1992). [f] TSCATS. [g] Platford, RF (1983).

Table 1. Structural formulas, chemical names, abbreviations and some properties of studied benzothiazoles

All benzothiazole used are solids at room temperature with the exception of benzothiazole (BT), which is liquid. Table 1 shows that BT possesses a high solubility (4300 mg/L), this is probably due to its high polarity and the fact that it is a liquid at room temperature. BT is also considered as volatile with a vapour pressure 0.0143 mm Hg (25°C).

4. MBT; most effective benzothiazoles for copper alloys inhibition

MBT is the most important member of the benzothiazole group of heterocyclic aromatic compounds. In fact, its discovery in ca. 1920 led to the major use in the production of rubber additive chemicals but predominately, as vulcanization accelerator in rubber industry.
MBT is also applied for various purposes, such as bio-corrosion inhibitor in industrial cooling and in the galvanic industry, and coating agent of metallic surfaces .It is also used as an external chemotherapeutic and antifungal drug in medical application .[16]
Both MBT and OHBT can exist in two tautomeric forms (Figure3).

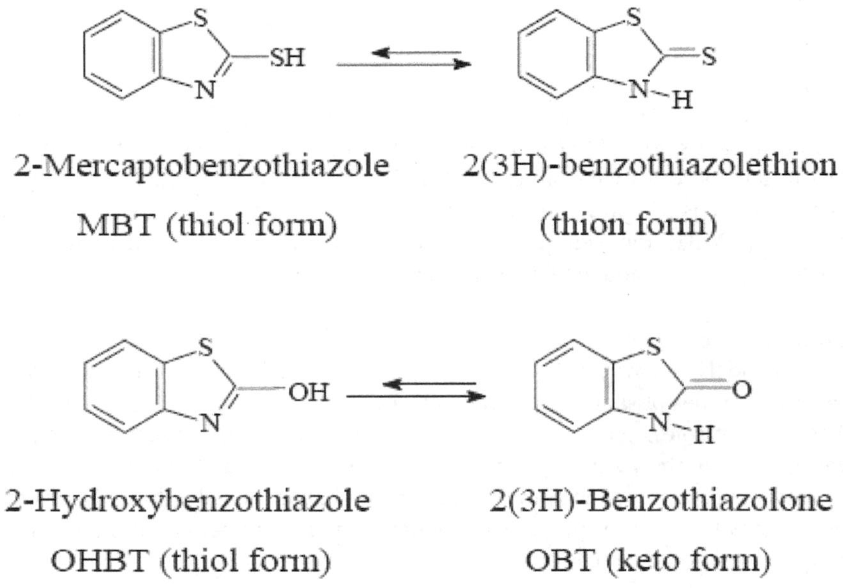

2-Mercaptobenzothiazole   2(3H)-benzothiazolethion

MBT (thiol form)   (thion form)

2-Hydroxybenzothiazole   2(3H)-Benzothiazolone

OHBT (thiol form)   OBT (keto form)

Fig. 3. Two tautomeric forms of MBT and OHBT

OHBT has a good solubility of 2354 mg/L in water. MBT, MTBT and TCMTB are moderately soluble with respective solubilities of 120 mg/L, 125 mg/L and 125 mg/L. These values of solubility and their respective vapour pressure (0.000464, 0.00026, and 3.12 10^{-7} mm Hg) show that these benzothiazoles are be considered as not volatile from aqueous solutions (or volatile with difficulty). [16]

4.1 Physical and chemical properties

The structural formula of 2-MBT is shown below:

(CAS Registry No: 149-30-4)

Technical 2-MBT is a yellowish to tan crystalline powder with a distinct, disagreeable odour. The solubility of 2-MBT in water under various conditions has been measured, as follows: 332 mg/L, pH unspecified; 51 mg/L at pH 5, 118 mg/L at pH 7, and 900 mg/L at pH 9; 120 mg/L at 24°C , 54 mg/L at 5°C , and 100-120 mg/L at 20°C . Solubility has also been measured in other solvents, including ethyl alcohol (20 g/L), acetone (100 g/L), benzene, and chloroform. 2-MBT has a specific gravity of 1.42-1.5 at 25°C, a vapour pressure of 24 mm Hg at 20°C, and a melting point of approximately 180°C. Decomposition occurs above 260°C. [17].

4.2 Toxicity

Some authors have shown that MBT is mainly responsible for toxic effects in MBT production activated sludge. MBT has been also shown to induce tumors, to be toxic (at 600 mmol L^{-1}) to aquatic organisms, and may also hamper waste treatment. MBT and MBTS have been reported as one of the most frequent allergens causing shoe dermatitis. Hinderer et al. proved that MBTS induced genetic damage to mammalian cells. MBT interfered with the nitrification processes and exhibited biocidal effects.[16]

4.3 Other applications

MBT is found widely in a variety of rubber articles in the modern environment both at home and at work. Examples of such articles are rubber tires and tubes for your car, rubber boots and shoes, rubber soles, gloves, garden hoses, elastic and rubberized clothing such as brassieres, girdles, support stockings, swimwear, swim caps and elastic bands as well as in rubber pillows, sponge makeup applicators, toys, balloons, baby bottle nippers, latex condoms, examination and surgical gloves, dental dams and rubber handles on tools such as tennis racquets and golf club handles. [18]

4.4 An effective corrosion iInhibitor for copper alloys

MBT is a particularly effective corrosion inhibitor for copper and copper alloy. In circulating cooling water system, low concentrations (such as 2 mg/L) of 2-Mercaptobenzothiazole (MBT) will be able to make copper and copper alloy corrosion rate dropped very low.

In direct currency cooling water system that the cooling equipment made of copper and copper alloy, because of the high usage and cost, MBT is rarely used as copper corrosion inhibitor.

Measurements of polarization curves show that the MBT at low concentrations is an anodic type inhibitor. MBT has many advantages: 1) effective corrosion inhibition control for copper and copper alloy; 2) low dosage. Shortcoming is very sensitive to chlorine and chloramines, it is easily destroyed by oxidation.

4.5 Analytical methods of detection

The Southern Research Institute used high pressure liquid chromatography (HPLC) to monitor the purity of radio labeled 2-MBT used in pharmacokinetic studies in mice and rats . Gradient elution with 20 mM acetic acid in 40% and 85% aqueous acetonitrile was used with radioactivity monitoring and UV absorbance at 254 nm. The more polar metabolites of 2-MBT in the urine were similarly analyzed except a combination of isocratic and gradient elution of 20 mM phosphoric acid in 20, 30, or 40% aqueous acetonitrile was also used. The detection limits of the method were not indicated in the available summaries.

A Japanese group[19] was able to achieve detection limits of 1.0 ug/g for fish tissues, and 10 ppb for water samples through extraction with methyl isobutylketone and analysis by HPLC.

Another Japanese group measured 2-MBT in water and sediment by extracting samples with methylene chloride and analyzing the extracts by gas-liquid chromatography using a flame photometric detector. Detection limits of 40 ppb for water and 2 ppb for sediment were achieved.[20]

Finally, Environment Canada has recently developed a liquid chromatography method for determining 2-MBT levels in effluents and sediments . The sample is extracted with methylene chloride, filtered, and concentrated. The residue is dissolved in acetonitrile and the sample is then analyzed for 2-MBT by HPLC. Detection limits are 25 ppb.[17]

The simple and convenient determination of 2-mercaptobenzothiazole (MBT) was spectrometrically performed with Cu complex in cationic CTAB media without an extraction procedure. This method has been studied in Tabriz Thermal Power Plant and we are reviewed details of this method in this part.

4.5.1 Spectrophotometric determination of 2-MBT in cooling water

4.5.1.1 Experimental

Instrumentation: A MiltonRoy 601(UV –Visible) spectrometer was used to measure the absorbance of Cu(II)-2-mercaptobenzothiazole complex in CTAB media. To adjust the pHs and prepare the buffer solution, Metrohm-827 pH meter was used.

Reagents and solutions: All chemicals, such as $CuSO_4$ (Riedel-de Haen) and 2-mercaptobenzothiazole (Accelerator), Methanol (Merck), Borax buffer (Merck) were analytical or guaranteed-grade reagents. Standard 2-MBT was made from 5.988 M stock solution. A 0.01% (w/v) cetyltrimethylammonium bromide (CTAB) (Merck) solution was prepared by dissolving 0.01 g of CTAB in a 100 mL volumetric flask with stirring; Cu(II) solution was prepared by dissolving in water to give a 0.005 M solution. Borax buffer (pH 9.0) was prepared by mixing 0.025 M borax and 0.1 M HCl.

Calibration curve: Standard 2-MBT solutions were prepared in range 2.9×10^{-6} M ~ 2.9×10^{-5} M. Several aliquots of 2-MBT standard solutions were taken in 50 mL volumetric flasks, and 2.0 mL of 0.01% CTAB and 1.0 mL of 0.005 M Cu(II) were added to each flask. Then it was filled to the mark with borax buffer solution (pH 9.0) and the calibration curve of 2-MBT was constructed by a UV-visible spectrophotometer. The regression equation was obtained with the method of least squares. Using this linear equation, we determined the correlation coefficient (R^2) and the detection limit. The detection limit is defined as the sample concentration giving a signal equal to the blank average signal plus three times the standard deviation of the blanks. [18].The calibration curve of Cu(II)-MBT complex with good linearity (R2=0.9995) was obtained at the concentration range between 2.9×10^{-6} and 2.9×10^{-5} M in 0.01% CTAB media. The detection limit was 9.7×10^{-7} M (0.162 mg L-1).

Application to real sample: The water of cooling system was taken as a real sample. The standard addition method was used to determine 2-MBT in real sample. A calibration curve was constructed at optimum conditions according to calibration curve procedure in Experimental Section. The calibration curve of Cu(II)-MBT complex with good linearity (R2 = 0.996) was obtained at the concentration range between 2.9×10^{-6} and 2.9×10^{-5} M in 0.01% CTAB media.

4.5.1.2 Results and discussion

Absorption spectra of Cu(II)-MBT complex: After Cu(II), MBT and CTAB were taken in a 50 mL volumetric flask so that their concentrations were 5×10^{-3} M and 1.2×10^{-5} M and 0.01%, respectively, the solution was diluted to the mark with borax buffer (pH 9.0). Then, the absorption spectrum of Cu(II)-MBT complex was obtained (Figure4). The analytical sensitivity and the reproducibility in this spectrum were good in CTAB media. The phenomenon seems to have been caused by the electrostatic and hydro-phobic interactions between Cu(II)-MBT complex and surfactant [21].

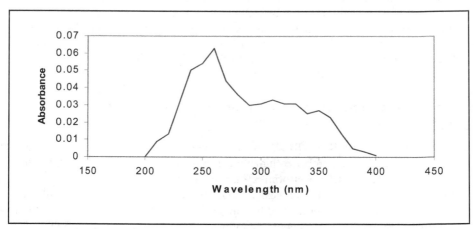

Fig. 4. UV-Visible spectra of Cu(II)-2- mercaptobenzothiazole (0.6×10^{-5} M) in 0.01% CTAB media at pH 9.0.

pH effect: The influence of pH on the absorbance of Cu(II)-MBT (0.6×10^{-5} M) complex in 0.01% CTAB media was investigated (Figure 5). Cu(II)-MBT complex showed the maximum absorption at pH 9.0. From this result, we realize that Cu(II)-MBT complex was quantitatively formed and well dissolved in CTAB media at pH 9.0. We assume that the reaction to form this complex could have competed against hydroxide precipitation above pH 9.0 and at acidic pH, as the sulfur atom in the chelating site of MBT has more affinity power with proton at a higher concentration of protons.

4.5.1.3 Conclusions

By using of Cu-MBT complex in CTAB bromide media, MBT could be determined simply, conveniently. Results from the proposed method shows that the calibration curve of Cu(II)-MBT complex with good linearity ($R2=0.9995$) was obtained at the concentration range between 2.9×10^{-6} and 2.9×10^{-5} M in 0.01% CTAB media. The detection limit was 9.7×10^{-7} M (0.162 mg L^{-1}). The proposed technique could be applied to the determination of MBT in real samples.

5. References

[1] A. Igual Muñoz, J. García Antón, J. L. Guiñón, V. Pérez Herranz, Electrochimica Acta 50, 957 (2004)

[2] J. B. Matos, L .P. Pereira, S. M. L.Agostinho, O. E. Barcia, G. G. O .Cordeiro, E. D'Elia, Journal of electroanalytical chemistry 570, 91 (2004)

[3] G. Moretti, F. Guidi, Corrosion science 44, 1995 (2002)

[4] E. M. Sherif, Su-Moon Park, Electrochimica Acta 51, 4665 (2006)

[5] E. M. Sherif, Su-Moon Park, Electrochem. Soc. 152, B428 (2005)

[6] E. Stupnisek-Lisac, A. Brnada, A. D. Mance, Corrosion science 42, 243 (2000)

[7] J. M. Bastidas, P. Pinilla, E. Cano, J. L. Polo, S. Miguel, Corrosion Science 45, 427 (2003)

[8] J. L. Polo, P. Pinilla, E. Cano, J. M. Bastidas, Corrosion, 59, 414 (2003)

[9] M. M. Singh, R. B. Rastogi, B. N. Upadhyay, M. Yadav, Materials chemistry and physics 80, 283 (2003)

[10] Marc Edwards, Loay Hidmi, Dawn Gladwell, Corrosion science 44, 1057 (2002)

[11] E. M. Sherif, Su-Moon Park, Electrochimica Acta 51, 6556 (2006)

[12] El-Sayed M. Sherif, Applied surface science 252, 8615 (2006)

[13] E. M. Sherif, Su-Moon Park, Corrosion science 48, 4065 (2006)

[14] El-Sayed M. Sherif, A. M. Shamy, Mostafa M. Ramla, Ahmed O. H. El Nazhawy, Materials chemistry and physics 102, 231 (2007)

[15] Corrosion inhibition in microelectronic copper thin film, national university of Singapore (2004)

[16] Microbial and photolytic degradation of benzothiazoles in water and wastewater, vorgelegte Dissertation von, Tag der wissenschaftlische Aussprache: 06. Juni (2003)

[17] H.W.Hanssen and N.D. Henderson, A review of the environment impact and toxic effects of 2-MBT, October 1991.23. Patient Information, MEKOS Laboratories (2005)

[18] Skoog, D. A.; Holler, F. J.; Nieman, T. A. Principles of Instrumental Analysis, 5th Ed.; Saunders College Publishing: Philadelphia, U.S.A., p 13, (1998)

[19] Ishiwata, A, Shimoda, Z, and Matsunaga, T. Determination of 2-mercaptobenzothiazole in fish and aquarium water. Nippon Gomu Kyokaishi: 51:813-822 (1978); Chem.Abstr. 90:34540(1979).

[20] Shinohara, J, Shinohara, R, Eto, S, and Hori, T. Micro Determinations of 2-mercaptobenzothiazole in water and sediment by gas chromatography with a flame photometric detector. Bunseki Kagaku 27:716-722(1978); Chem. Abstr. 90:66263(1979).

[21] Esteve-Romero, J. S.; Monferrer-Pons, L.; Ramis-Ramos, G.;Garcia-Alvarez-Coque, M. C. Talanta, 42, 737 (1995)

[22] Spacu, G.; Kuras, M. Z. Anal. Chem., 102, 108 (1936)

Permissions

The contributors of this book come from diverse backgrounds, making this book a truly international effort. This book will bring forth new frontiers with its revolutionizing research information and detailed analysis of the nascent developments around the world.

We would like to thank Associate Professor Mohammad Rasul (Mechanical Engineering), for lending his expertise to make the book truly unique. He has played a crucial role in the development of this book. Without his invaluable contribution this book wouldn't have been possible. He has made vital efforts to compile up to date information on the varied aspects of this subject to make this book a valuable addition to the collection of many professionals and students.

This book was conceptualized with the vision of imparting up-to-date information and advanced data in this field. To ensure the same, a matchless editorial board was set up. Every individual on the board went through rigorous rounds of assessment to prove their worth. After which they invested a large part of their time researching and compiling the most relevant data for our readers. Conferences and sessions were held from time to time between the editorial board and the contributing authors to present the data in the most comprehensible form. The editorial team has worked tirelessly to provide valuable and valid information to help people across the globe.

Every chapter published in this book has been scrutinized by our experts. Their significance has been extensively debated. The topics covered herein carry significant findings which will fuel the growth of the discipline. They may even be implemented as practical applications or may be referred to as a beginning point for another development. Chapters in this book were first published by InTech; hereby published with permission under the Creative Commons Attribution License or equivalent.

The editorial board has been involved in producing this book since its inception. They have spent rigorous hours researching and exploring the diverse topics which have resulted in the successful publishing of this book. They have passed on their knowledge of decades through this book. To expedite this challenging task, the publisher supported the team at every step. A small team of assistant editors was also appointed to further simplify the editing procedure and attain best results for the readers.

Our editorial team has been hand-picked from every corner of the world. Their multi-ethnicity adds dynamic inputs to the discussions which result in innovative outcomes. These outcomes are then further discussed with the researchers and contributors who give their valuable feedback and opinion regarding the same. The feedback is then collaborated with the researches and they are edited in a comprehensive manner to aid the understanding of the subject.

Apart from the editorial board, the designing team has also invested a significant amount of their time in understanding the subject and creating the most relevant covers. They scrutinized every image to scout for the most suitable representation of the subject and create an appropriate cover for the book.

The publishing team has been involved in this book since its early stages. They were actively engaged in every process, be it collecting the data, connecting with the contributors or procuring relevant information. The team has been an ardent support to the editorial, designing and production team. Their endless efforts to recruit the best for this project, has resulted in the accomplishment of this book. They are a veteran in the field of academics and their pool of knowledge is as vast as their experience in printing. Their expertise and guidance has proved useful at every step. Their uncompromising quality standards have made this book an exceptional effort. Their encouragement from time to time has been an inspiration for everyone.

The publisher and the editorial board hope that this book will prove to be a valuable piece of knowledge for researchers, students, practitioners and scholars across the globe.

List of Contributors

Liping Li
China Power Engineering Consulting (Group) Corporation, China

Eric Hu
School of Mechanical Engineering, the University of Adelaide, Australia

Yongping Yang
North China Electric Power University, Beijing, China

Akira Nishimura
Division of Mechanical Engineering, Mie University, Tsu, Japan

Masayuki Taniguchi
Hitachi Research Laboratory, Hitachi, Ltd., 7-1-1 Omika-cho, Hitachi-shi, Ibaraki-ken, Japan

Hamdi Mohamed and Benticha Hmaeid
Laboratoire d'Etudes des Systèmes Thermiques et Energétique, Tunisia

Sassi Mohamed
Masdar Institute of Science and Technology, Abu Dhabi, United Arab Emirates

Ahmed Mahmoud Hegazy
Engineering for Petroleum and Process Industries (Enppi), Cairo, Egypt

M.N. Lakhoua
Member IEEE, Laboratory of Analysis and Command of Systems (LACS), ENIT, Le Belvedere, Tunis, Tunisia

E. A. Ogbonnaya and K. T. Johnson
Department of Marine Engineering, Rivers State University of Science and Technology, Port Harcourt

H. U. Ugwu
Department of Mechanical Engineering, Michael Okpara University of Agriculture, Umudike-Umuahia

C. A. N. Johnson
Department of Marine Engineering, Niger Delta University, Wilberforce Island, Bayelsa State

Barugu Peter Forsman
Department of Welding, Oil and Gas Engineering, Petroleum Training Institute, Effurun, Delta State, Nigeria

Ryszard Szczepanik, Radosław Przysowa, Jarosław Spychała, Edward Rokicki, Krzysztof Kaźmierczak and Paweł Majewski
Instytut Techniczny Wojsk Lotniczych (ITWL, Air Force Institute of Technology), Poland

Soner Gokten
Gazi University, Turkey

S. Moazzem, M.G. Rasul and M.M.K. Khan
School of Engineering and Built Environment, Faculty of Sciences, Engineering and Health, Central Queensland University, Rockhampton, Queensland, Australia

Zheng-Fei Hu
School of Materials Science and Engineering, Tongji University, Shanghai, China

Fazael Mosaferi, Farid Delijani and Fateme Ekhtiary Koshky
East Azarbayjan Power Generation, Management Company, Tabriz, Iran

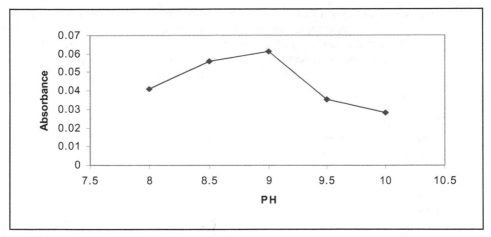

Fig. 5. Effect of pH on the absorbance of Cu(II)-2-mercaptobenzothiazole(0.6×10^{-5} M) in 0.01% CTAB media

Concentration of CTAB: When the concentration of CTAB surfactant exceeds its critical micelle concentration, the homogeneous micelle solution is formed at a point where Cu(II)-MBT complex can be well dissolved. Due to high viscosity, the concentrated CTAB media was hard to handle, whereas those with low viscosity under diluted conditions could not form a micelle or make a homogeneous solution of complex as the polarity of aqueous solution was not lowered. With the concentration of CTAB varying from 0.005% to 0.03% at pH 9.0, the absorbance of Cu(II)-MBT (0.6×10^{-5}M) complex was investigated and the results are shown in Figure 6. The maximum absorbance was obtained when the concentration of CTAB was 0.01%.

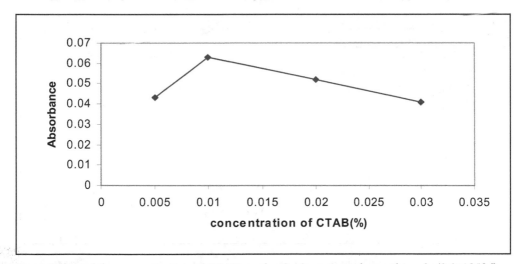

Fig. 6. Effect of the concentration of CTAB on the Cu-2-mercaptobenzothiazole (0.6×10^{-5} M) complex at PH 9.

Concentration of Cu: It is known that Cu(II) is stochiometrically combines with MBT to form 1 : 2 complex [22]. For a metal complex to be formed quantitatively, however, one must add more chelating agent to the sample solution. Figure 7 shows how the absorbance of Cu(II)-MBT complex changes with the concentration of Cu. We found that when Cu was added to more than 125 equivalent of MBT (to the mol), the absorbance was high and constant.

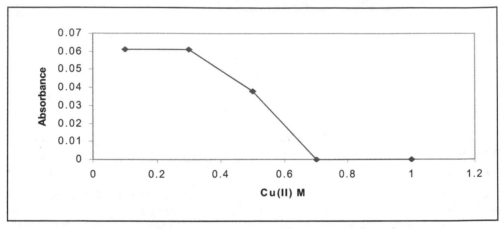

Fig. 7. Effect of the concentration of Cu on the Cu-2-mercaptobenzothiazole(0.6×10^{-5}M) complex at PH 9.

To investigate the stability of Cu(II)-MBT complex in CTAB media at pH 9.0, the absorbance was measured as the function of time (Figure 8). The absorbance is constant from the beginning of measurement to 20 min and after 20 min, the absorbance was decreased.

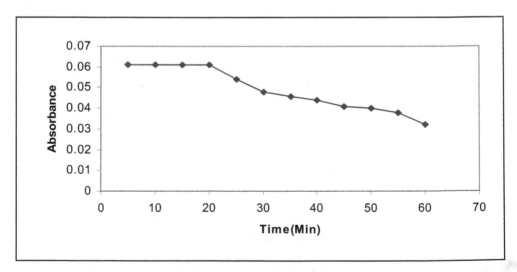

Fig. 8. Effect of time on the stability Cu-2-mercaptobenzothiazole (0.6×10^{-5} M) complex at PH 9.

Printed in the USA
CPSIA information can be obtained
at www.ICGtesting.com
JSHW011450221024
72173JS00004B/1015

9 781632 404049